ROUTLEDGE LIBRARY EDTIONS: GLOBAL TRANSPORT PLANNING

Volume 19

TRANSPORTATION NETWORKS

TRANSPORTATION NETWORKS
A Quantitative Treatment

D. TEODOROVIĆ

Routledge
Taylor & Francis Group
LONDON AND NEW YORK

First published in 1986 by Gordon and Breach, Science Publishers

This edition first published in 2021
by Routledge
2 Park Square, Milton Park, Abingdon, Oxon OX14 4RN

and by Routledge
52 Vanderbilt Avenue, New York, NY 10017

Routledge is an imprint of the Taylor & Francis Group, an informa business

© 1986 OPA (Amsterdam) B.V.

All rights reserved. No part of this book may be reprinted or reproduced or utilised in any form or by any electronic, mechanical, or other means, now known or hereafter invented, including photocopying and recording, or in any information storage or retrieval system, without permission in writing from the publishers.

Trademark notice: Product or corporate names may be trademarks or registered trademarks, and are used only for identification and explanation without intent to infringe.

British Library Cataloguing in Publication Data
A catalogue record for this book is available from the British Library

ISBN: 978-0-367-69870-6 (Set)
ISBN: 978-1-00-316032-8 (Set) (ebk)
ISBN: 978-0-367-74703-9 (Volume 19) (hbk)
ISBN: 978-0-367-74712-1 (Volume 19) (pbk)
ISBN: 978-1-00-315916-2 (Volume 19) (ebk)

Publisher's Note
The publisher has gone to great lengths to ensure the quality of this reprint but points out that some imperfections in the original copies may be apparent.

Disclaimer
The publisher has made every effort to trace copyright holders and would welcome correspondence from those they have been unable to trace.

TRANSPORTATION NETWORKS
A Quantitative Treatment

D. Teodorović
University of Belgrade

GORDON AND BREACH SCIENCE PUBLISHERS
New York · London · Paris · Montreux · Tokyo

© 1986 by OPA (Amsterdam) B.V. All rights reserved.
Published under license by Gordon and Breach
Science Publishers S.A.

Gordon and Breach Science Publishers

P.O. Box 786
Cooper Station
New York, NY 10276
United States of America

P.O. Box 197
London WC2E 9PX
England

58, rue Lhomond
75005 Paris
France

P.O. Box 161
1820 Montreux 2
Switzerland

14-9 Okubo 3-chome,
Shinjuku-ku
Tokyo 160
Japan

Library of Congress Cataloging-in-Publication Data
Teodorovic, D. (Dusan), 1951–
  Transportation networks.

  (Transportation studies, 0278-3819; v. 6)
  Translated from Serbo-Croatian (roman).
  Includes index.
  1. Transportation—Planning—Mathematical models.
2. Network analysis (Planning) I. Title. II. Series.
HE147.7.T46 1985   380.5′068   85-27302
ISBN 0-677-21380-8

ISBN 0-677-21380-8. ISSN 0278-3819. No part of this book may be reproduced or utilized in any form or by any means, electronic or mechanical, including photocopying and recording, or by any information storage or retrieval system, without permission in writing from the publishers. Printed in Great Britain by Bell and Bain Ltd., Glasgow.

Contents

Introduction to the Series		vii
Preface		ix
1	**Shortest Paths in Transportation Networks**	1
	1.1 Introduction	1
	1.2 Basic Concepts Behind the Theory of Transportation Networks	2
	1.3 Method for Finding the Shortest Path from One Node to all Other Nodes in the Transportation Network	10
	1.4 Method for Finding the Shortest Path Between all Pairs of Nodes	21
	1.5 Method for Determining Shortest Paths When the Transportation Network Contains Two Types of Branches	29
	1.6 Finding the Shortest Path in a Probabilistic Network	34
	1.7 Method for Finding Minimum Spanning Trees	49
2	**Transportation Network Flows**	49
	2.1 Flows on Transportation Networks	49
	2.2 Conservation Laws on Transportation Networks	50
	2.3 Flows on Transportation Networks with One Source and One Sink	53
	2.4 Branch Capacities and Transportation Network Capacities	56
	2.5 Algorithm for Finding the Maximum Flow Through a Transportation Network	57

3	**Vehicle Routing Problems on Networks**	**65**
3.1	Vehicle Routing on the Network	68
3.2	The Chinese Postman's Problem on a Nonoriented Network	69
3.3	The Chinese Postman Problem on an Oriented Network	78
3.4	The Travelling Salesman Problem	86
3.5	Designing Optimal Routes for Vehicles Which Must go Through all Nodes of the Transportation Network at Least Once When There are Time Constraints or Vehicle Capacity Constraints	114
3.6	Heuristic Algorithm for the Traveling Salesman Problem	118
3.7	Routing and Scheduling Problems for Several Vehicles on the Transportation Network	124
3.8	Classification of Vehicle Routing and Scheduling Problems on a Transportation Network	125
3.9	Designing Optimal Routes for a Fleet of Vehicles Which Must Service Every Node on the Transportation Network	128
3.10	The Problem of Routing Several Vehicles When There is Only One Depot	131
3.11	Heuristic "Sweeping" Algorithm to Route Vehicles on a Transportation Network When There is One Depot	139
3.12	Heuristic "Assignment" Algorithm to Route Vehicles on a Transportation Network When There is One Depot	142
3.13	The Problem of Vehicle Routing When There are Several Depots	150
3.14	Method to Determine the Minimum Number of Vehicles Needed to Service a Given Schedule on the Transportation Network	155
3.15	Optimal Dispatching Strategy on a Transportation Network After a Schedule Perturbation	172

4	**Determining Vehicle Depot Locations**		185
	4.1	Determining Transportation Network Centers	186
	4.2	The Problem of Several Centers	192
	4.3	The Median Problem	193
	4.4	Algorithm to Determine One Network Median	195
	4.5	Algorithm for Finding k Medians	198

Introduction to the Series

This broad-ranging series of books will cover many of the varied aspects of transportation. The subject area will be generally divided into two parts: the first deals with planning and technological aspects of transportation, the second with specialized transportation.

Within the context of this series, technology and planning will include the wide spectrum of various aspects of the design and planning of vehicles and infrastructure for the transport of freight and passengers, as well as operational and management considerations. The general aim of the planning and technology series is to inform readers of the state of the art and to summarize the status of transportation.

The second part of the series will seek to generate monographs dealing with improving the mobility of those groups in society increasingly characterized as the transportation disadvantaged, particularly, but not exclusively, the elderly, the disabled and families with low incomes. It is anticipated that the content of these books will be derived largely from research, policy analysis and documental field experience. The subject matter will include advances in the relevant technology, service and methods demonstrations, improved planning and methodology, major or proposed changes in public policy and innovative proposals for the development or change of systems.

Occasionally, more specific monographs will be published, presenting the results of individual studies in areas of special interest to planners and technologists.

As with any monograph series, the emphasis is on current information, and the material will be of interest to the transport practitioner, the postgraduate student and the academic working in the field.

NORMAN ASHFORD
WILLIAM G. BELL

Preface

In recent years, a large number of papers have been published throughout the world from the field of transportation networks. This book is the author's attempt to contribute to the further study of this subject. It is primarily meant for those who are encountering transportation network problems for the first time. During the course of writing, the simplest possible mathematical methods were utilized to facilitate comprehension.

I would particularly like to thank Alice Copple Tošić for her exceptional support in translating and preparing the manuscript for printing.

D. TEODOROVIĆ

1 Shortest Paths in Transportation Networks

1.1. Introduction

A transportation network can be defined as a set of nodes and a set of branches on which transportation activities are carried out. These networks are encountered in all fields of traffic and transportation. Depending on the area of transportation involved, nodes can signify cities, street crossings, airports, train stations, quays, bus stations, freight terminals, etc. Nodes in a transportation network are linked by specific branches which can be denoted by streets, roads, air routes, railroad tracks or waterways. On all transportation networks requests arise in certain nodes to transport goods or passengers or to transmit certain information.

Infrastructural costs, costs related to providing transportation, and the level of service all basically depend on the transportation network's design and on the organization of transportation on the network.

The theory of transportation networks as a general traffic and transportation discipline has been increasingly developed in the world in recent times and can make an essential contribution to improved operations of an entire transportation system. The application of existing theoretical models, modifying them and creating new ones can achieve considerable economic effects and improve the level of service.

1.2. Basic concepts behind the theory of transportation networks

Transportation networks are denoted in the same manner as graphs. Although the terms graph and network are used interchangeably, we make a distinction between them to the effect that <u>graph</u> denotes the structural relationship between nodes and <u>network</u> refers to a graph which has quantitative relationships between branches and nodes.

The symbols which specify transportation network G (N,A) designate a network containing a set of nodes denoted by N and a set of links between these nodes denoted by A. Links in the transportation network are often called branches. The notation (i,j) denotes the link or branch which connects node $i \in N$ with node $j \in N$.

If all links or branches in the transportation network are oriented, the network is called an oriented network. In an oriented network, branch (i,j) leads from node i to node j. In the opposite case when none of the branches are oriented, the network is called nonoriented. If some of the branches in the network are oriented and some nonoriented, this is called a mixed network. Figures 1, 2 and 3 provide examples of oriented, nonoriented and mixed networks, respectively.

SHORTEST PATHS IN TRANSPORTATION NETWORKS   3

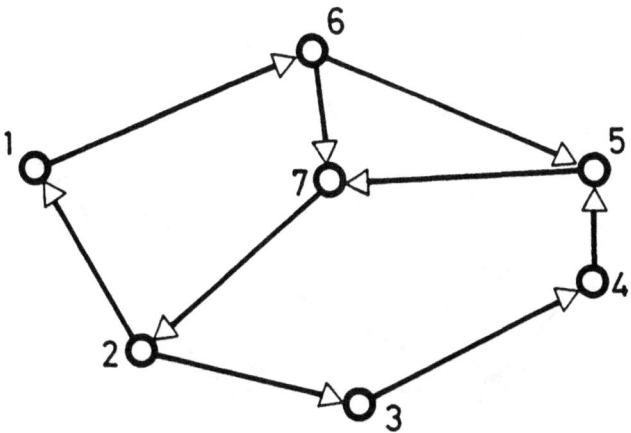

FIGURE 1. An oriented network.

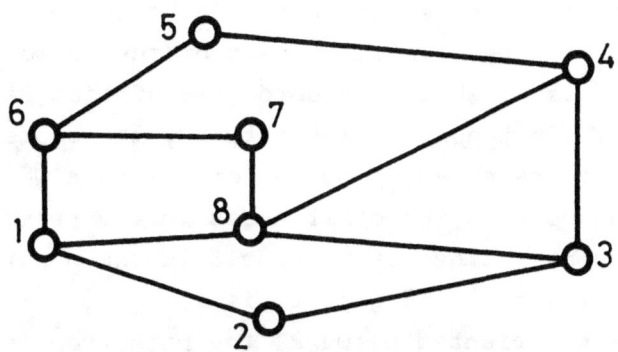

FIGURE 2. A nonoriented network.

4 TRANSPORTATION NETWORKS

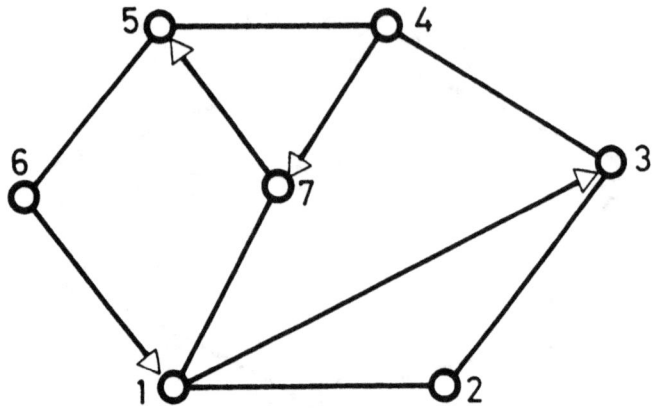

FIGURE 3. A mixed network

The indegree of a node in an oriented network represents the number of branches which enter that node. Analogous to this, the outdegree of a node in an oriented network is defined as the number of branches which leave that node. For example, the indegree of node 5 which is part of the network in Figure 1, is 2, while the outdegree of that same node is 1. In a nonoriented network, the degree of the node represents the number of branches which link that node to the other nodes in the network. For example, the degree of node 8 in the nonoriented network shown in Figure 2, is 4.

In a nonoriented network, the path from node i to node j includes all those branches and nodes which must be passed when going from node i to node j. The path is defined either by counting the nodes or by counting the branches which are passed. For example, $(1,6,7)$ denotes the path which leads from node 1 through node 6 to node 7 (Figure 2). This path can also be denoted by $((1,6), (6,7))$ which means that the path from node 1 to node 7 passes through branches (1,6) and (6,7).

A cycle signifies a path which starts and finishes at the same node. For example, $(1,8,7,6,1)$ describes the cycle, or path, which starts and finishes at node 1 (Figure 2). The paths in an oriented network are most often called chains. The starting and finishing nodes on the oath are called the source node and the destination node. If individual branches only appear on the path once, the path is called a simple path. The path is called elementary if each node appears on the path only once.

Nodes i and j are said to be connected if there is a path leading from node i to node j. In accordance with this, a nonoriented network is said to be connected if there is a corresponding path between all pairs of nodes i, j$\in$N.

An oriented network is said to be connected if the corresponding nonoriented network (which is obtained when the orientations are removed) is connected. It is clear that an oriented connected network does not have paths between all pairs of nodes due to its oriented character. An oriented network which has paths between all pairs of nodes is called a strongly connected oriented network.

Figure 4 shows a nonoriented connected network and Figure 5 shows the same network with oriented branches.

6   TRANSPORTATION NETWORKS

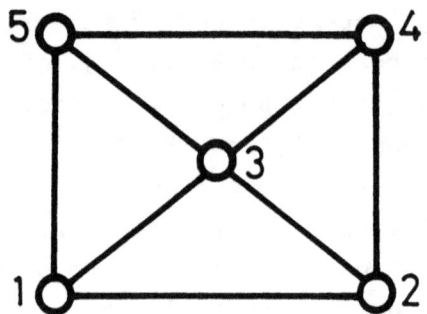

FIGURE 4. A nonoriented connected network.

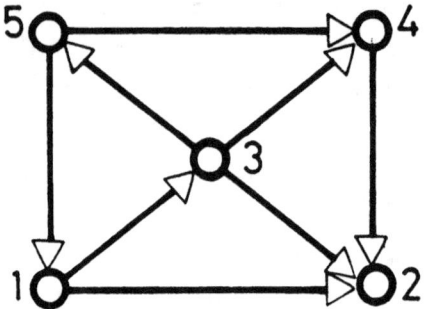

FIGURE 5. An oriented connected network.

The network on Figure 5 is an oriented connected network since the nonoriented network from which it arose (the network on Figure 4) is connected and since paths do not exist between all ordered pairs of nodes. For example, it is not possible to go from node 4 to node 5, to node 3 or to node 1.

Figure 6 shows a strongly connected oriented network.

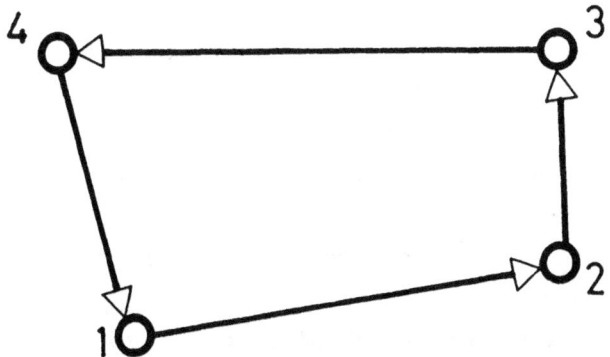

FIGURE 6. A strongly connected oriented network.

Oriented graph G_1 (N, A_1) ia a partial graph of oriented graph G (N, A) for which A_1 is a subset of A, i.e. in which $A_1 \subset A$. Oriented graph G_2 (N_2, A_2) is a subgraph of graph G (N, A) so that $N_2 \subset N$ and $A_2 \subset A$. This means that only branches which connect the nodes of set N_2 belong to the set of branches A_2 in subgraph G_2, so that :

$$A_2 = \left[(i,j) \ \middle| \ (i,j) \in A, \quad i \in N_2, \ j \in N_2\right]$$

The partial graph of an oriented graph can be obtained by eliminating certain branches of an oriented graph. A subgraph is obtained by eliminating certain nodes and incidental branches (branches which start or finish in the nodes which are eliminated).

In both the theory of graphs and the theory of transportation networks, the concept of trees is particularly interesting. The tree of a nonoriented network is comprised of a connected subgraph which does not have any cycles. Since a tree is above all a subgraph, this means that it only contains a certain

number of nodes from the original network. And
because a tree is defined as a connected subgraph,
we can conclude that a corresponding path exists
between all pairs of nodes on the tree. Finally,
since we said that a tree does not contain any
cycles, it can be concluded that none of the tree's
paths starts and finishes in the same node.

A spanning tree comprises all the nodes of the
original network.

A tree originating from an oriented network is
called an oriented tree. This type of tree chara-
teristically contains a root node from which a
unique path leads to any other node in the tree.
It is clear that an oriented tree can also contain
all the nodes of the original network.

Figures 7 and 8 show a nonoriented network and
one tree which originated from this network.

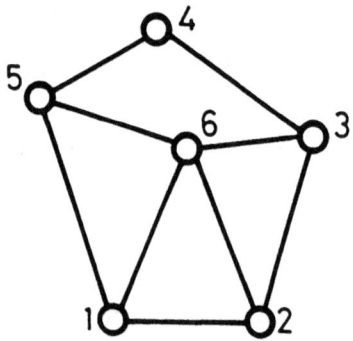

FIGURE 7. A nonorient-
ed network.

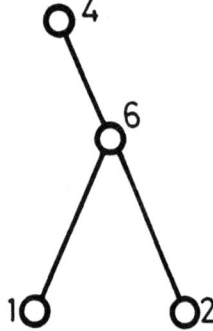

FIGURE 8. A tree origi-
nating from the network
in Fig. 7.

Figures 9 and 10 give another example of a network and a spanning tree arising from the network.

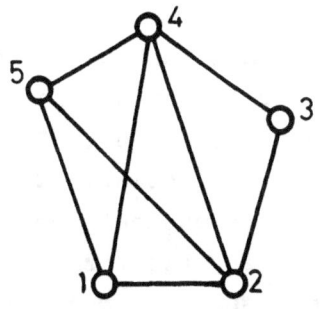 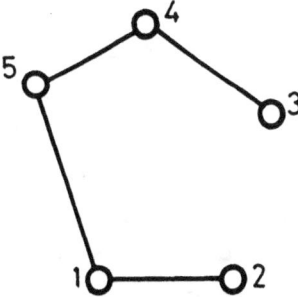

FIGURE 9. A nonoriented network.

FIGURE 10. A spanning tree from the network in Fig. 9.

An oriented network and an oriented tree arising from it containing a root node are shown in Figures 11 and 12.

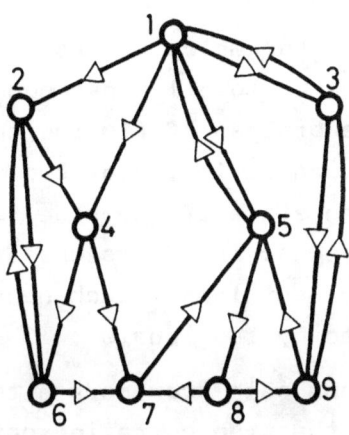 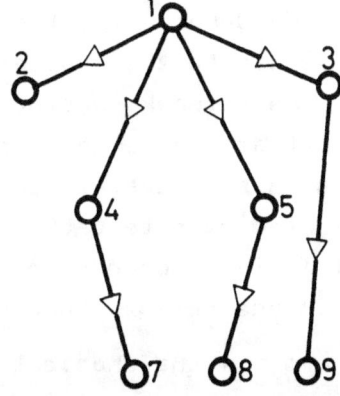

FIGURE 11. An oriented network.

FIGURE 12. An oriented tree with root node.

1.3. Method for finding the shortest path from one node to all other nodes in the transportation network

One of the oldest problems in both the theory of transportation networks and the theory of graphs is the problem of finding the shortest path from one node to all other nodes in the network. Numerous papers have been published throughout the world dealing with this problem and many algorithms have been created to determine this shortest path. Some of these were later shown to be rather inefficient, while others were actually modifications of several of the best-known algorithms. One of the most efficient algorithms for finding the shortest path from one node to all other nodes in a network was given by E.W. Dijkstra in 1959[15] as follows .

This algorithm assumes above all that the lengths of all branches $l(i,j)$ in network $G(N,A)$ are nonnegative.

We denote by a the node for which we are to investigate the shortest paths to all other nodes in the network. During the process of finding these shortest paths, each node can be in one of two possible states, in an open state if the node is denoted by a tentative label or in a closed state if it is denoted by a permanent label. Each node i in the network is denoted by two labels :

d_{ai} - the shortest known path from node a to node i found in the transportation network so far,

q_i – the node located in front of node i on the shortest known path from node a to node i found so far in the transportation network.

The last node which we have established as being in a closed state in the process is denoted by c. The symbol + denotes the immediate predecessor node of node a.

The Dijkstra algorithm is comprised of the following 5 steps :

Step 1 : The process starts from node a. Since the length of the shortest path from node a to node a is 0, then d_{aa} = 0. We stated above that the immediate predecessor node of node a will be denoted by the symbol + so that q_a = +. Since the lengths of the shortest paths from node a to all other nodes i ≠ a are for the present uninvestigated, we tentatively put d_{ai} = ∞ for i ≠ a. Since all immediate predecessor nodes to nodes i ≠ a on the shortest path are unknown, we put q_i = – for all i ≠ a. The only node which is now in a closed state is node a. Therefore we write that c = a.

Step 2 : In order to transform some of the temporary labels into permanent labels, we examine all branches (c,i) which exit from the last node which is in a closed state (node c). If node i is also in a closed state, we pass the examination on to the next node. If node i is in an open state, we obtain its first label d_{ai} based on the equation :

$$d_{ai} = \min \left[d_{ai}, d_{ac} + l(c,i) \right]$$

in which the left side of the equation is the new label of node i. We should note that d_{ai} appearing on the right side of the equation is the old label for node i.

Step 3: In order to determine which node will be the next to go from an open to a closed state, we compare value d_{ai} for all nodes which are in an open state and choose the node with the smallest d_{ai}. Let this be some node j. Node j passes from an open to a closed state since there is no path from a to j shorter than d_{aj}. The path through any other node would be longer.

Step 4: We have ascertained that j is the next node to pass from an open state to a closed one. We then determine the immediate predecessor node of node j and the shortest path which leads from node a to node j. We examine the length of all branches (i,j) which lead from closed state nodes to node j until we establish that the following equation is satisfied :

$$d_{aj} - l(i,j) = d_{ai}$$

Let this equation be satisfied for some node t. This means that node t is the immediate predecessor of node j on the shortest path which leads from node a to node j. Therefore, we can write that $q_j = t$.

Step 5: Node j is in a closed state. When all nodes in the network are in a closed state, we have completed the process of finding the shortest path. Should any node still be in an open state, we return to Step 2.

The algorithm described above can also be used to find the shortest path between two specific nodes. In this case, the algorithm is completed when both nodes are in a closed state. This algorithm can be applied to find the shortest path in nonoriented, oriented and mixed networks.

E x a m p l e : Using the Dijkstra algorithm, calculate the shortest path from node a to all other nodes on the transportation network shown in Figure 13.

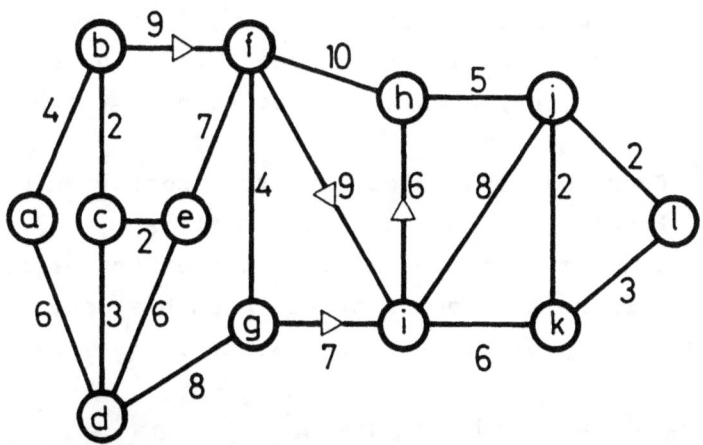

FIGURE 13.   Transportation network G(N,A).

14 TRANSPORTATION NETWORKS

The numbers next to the network branches shown on Fig. 13 signify the length of each branch. (We should mention that the length of a branch in a network could also, in addition to actual length, stand for travel time or travel expenses or numerous other values).

We start the process of finding the shortest path at node a. Since the length of the shortest path from node a to node a is 0, then $d_{aa} = 0$. The immediate predecessor node of starting node a is denoted by the symbol + so that $q_a = +$. The length of all other shortest paths from node a to all other nodes $i \neq a$ are for the present unexamined, so for all other nodes $i \neq a$ we put $d_{ai} = \infty$. Since the immediate predecessor nodes of nodes $i \neq a$ on the shortest path are unknown, we put $q_i = -$ for all $i \neq a$. The only node which is now in a closed state is node a. Therefore, c = a. Next to the node a symbol we put the label (0,+) and add the symbol ' to emphasize that node a is in a closed state. This completes the first step of the algorithm.

After the first step, the transportation network looks like Fig. 14 on the next page.

We now move to the second step of the algorithm. By examining the length of all branches leaving node a which are in a closed state we can write :

$$d_{ab} = \min\left[\infty, 0 + 4\right] \qquad d_{ab} = 4$$

$$d_{ad} = \min\left[\infty, 0 + 6\right] \qquad d_{ad} = 6$$

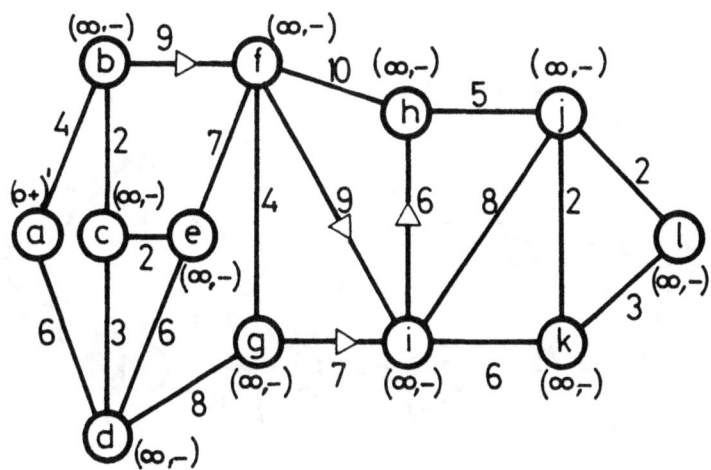

FIGURE 14. Network G(N,A) after the first time through the algorithm.

In step 3 we determine which node will be next in line to pass from an open to a closed state. Since $d_{ab} < d_{ad}$ node b passes from an open to a closed state. In the same manner, since :

$$d_{ab} - l\,(a,b) = 4 - 4 = 0 = d_{aa}$$

we can conclude in step 4 that node a is the immediate predecessor of node b on the shortest path, i.e. $q_b = a$.

Now in step 5 we note that there are still many nodes which are in an open state. Therefore, we have to return to step 2. The transportation network looks like this after going through all five steps of the algorithm for the first time :

16   TRANSPORTATION NETWORKS

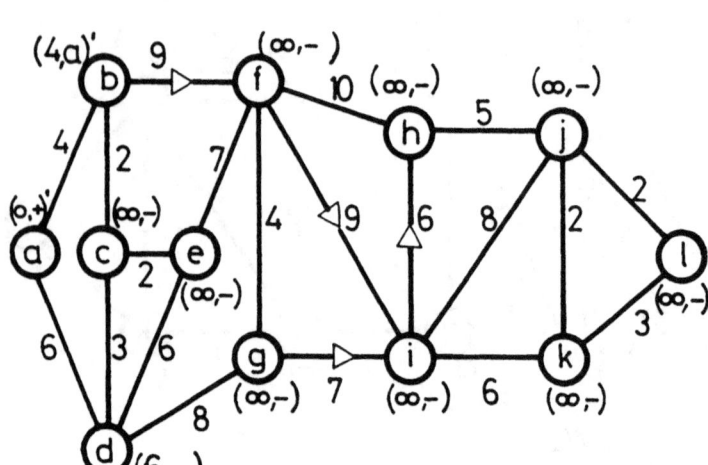

FIGURE 15.   Transportation network G(N,A) after first time through all 5 steps of the algorithm.

Let us now return to the second step of the algorithm. The last node to go from an open to a closed state was node b. This means that c = b. When we examine all branches leaving node b going towards nodes which are in an open state, we have :

$$d_{af} = \min \left[\infty, d_{ab} + 1\ (b,f) \right] =$$
$$= \min \left[\infty, 4 + 9 \right] = \min \left[\infty, 13 \right] = 13$$

$$d_{ac} = \min \left[\infty, d_{ab} + 1\ (b,c) \right] =$$
$$= \min \left[\infty, 4 + 2 \right] = \left[\infty, 6 \right] = 6$$

Since $d_{ac} < d_{af}$ node c is the next to switch from an open to a closed state. We then determine the immediate predecessor node of node c. Since :

SHORTEST PATHS IN TRANSPORTATION NETWORKS   17

$$d_{ac} - 1\,(b,c) = 6 - 2 = 4 = d_{ab}$$

then node b is the immediate predecessor of node c
and $q_c = b$.

The network still contains many nodes in an
open state, so we must once again return to step 2
of the algorithm. After the second time through all
5 steps of the algorithm, the transportation network
looks like this :

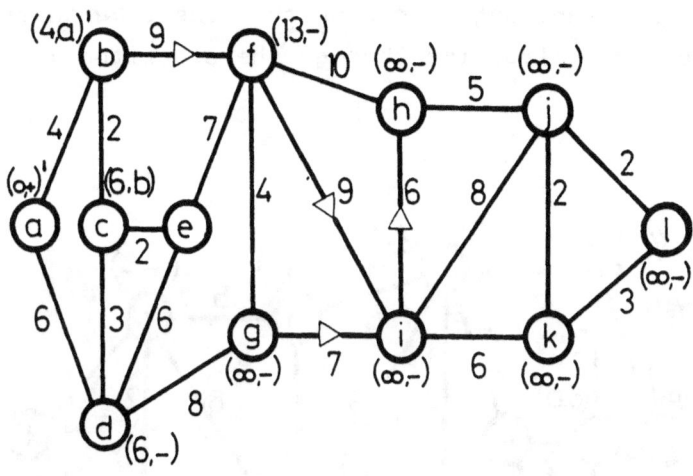

FIGURE 16. Network G(N,A) **second** time through
the algorithm's 5 steps.

The third time through the algorithm gives
us :

$$d_{ae} = \min\left[\infty,\; d_{ac} + 1\,(c,e)\right] =$$
$$= \min\left[\infty,\; 6 + 2\right] = \min\left[\infty,\; 8\right] = 8$$

18 TRANSPORTATION NETWORKS

$$d_{ad} = \min \left[6, d_{ac} + 1(c,d) \right] =$$
$$= \min \left[6, 6 + 3 \right] = \min \left[6, 9 \right] = 6$$

Since $d_{ad} < d_{ae}$ node d is the next node to switch to a closed state. Since :
$$d_{ad} - 1(a,d) = 6 - 6 = 0 = d_{aa}$$
then node a is the immediate predecessor of node d on the shortest path, so $q_d = a$. And now c = d.

The transportation network looks like this after the third time through the algorithm :

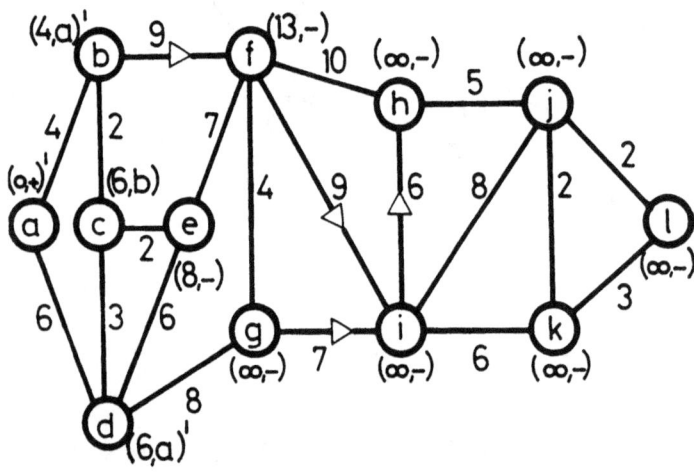

FIGURE 17. Network G(N,A) third time through the algorithmic steps.

SHORTEST PATHS IN TRANSPORTATION NETWORKS   19

Table I on the following page shows the calculations leading to the shortest paths and determination of the immediate predecessor nodes after going through the algorithm 11 times.

The shortest paths from node a to all other nodes in the network are shown on Fig. 18 with immediate predecessor nodes noted after going through the algorithm 11 times.

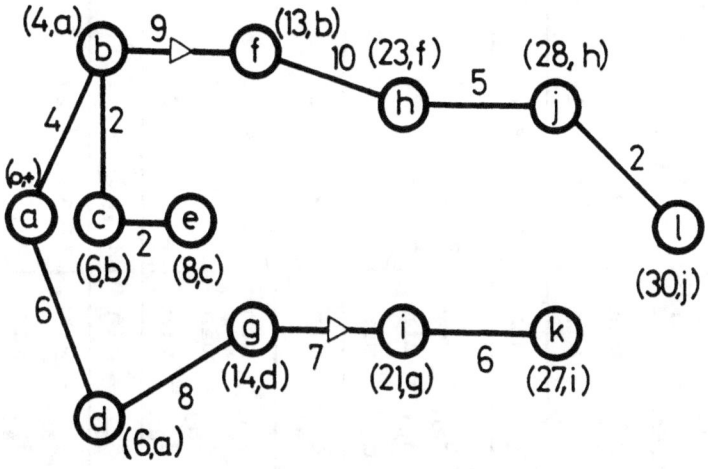

FIGURE 18. Shortest path from node a to all other nodes in network G(N,A).

20 TRANSPORTATION NETWORKS

TABLE I Calculating the shortest paths and predecessor nodes

Times through algorithm	Last node in closed state	Branches from last closed node to open nodes $d_{ai} = \min[d_{ai}, d_{ac}+1(c,i)]$	Next node in closed state	Predecessor node
4	d	$d_{ae} = \min\begin{bmatrix}8,12\end{bmatrix} = 8$ $d_{ag} = \min\begin{bmatrix}\infty,14\end{bmatrix} = 14$	e	$q_e = c$
5	e	$d_{af} = \min\begin{bmatrix}13,15\end{bmatrix} = 13$	f	$q_f = b$
6	f	$d_{ah} = \min\begin{bmatrix}\infty,23\end{bmatrix} = 23$ $d_{ai} = \min\begin{bmatrix}\infty,22\end{bmatrix} = 22$ $d_{ag} = \min\begin{bmatrix}14,17\end{bmatrix} = 14$	g	$q_g = d$
7	g	$d_{ai} = \min\begin{bmatrix}22,21\end{bmatrix} = 21$	i	$q_i = g$
8	i	$d_{ah} = \min\begin{bmatrix}23,27\end{bmatrix} = 23$ $d_{aj} = \min\begin{bmatrix}\infty,29\end{bmatrix} = 29$ $d_{ah} = \min\begin{bmatrix}\infty,27\end{bmatrix} = 27$	h	$q_h = f$
9	h	$d_{aj} = \min\begin{bmatrix}29,28\end{bmatrix} = 28$	j	$q_j = h$
10	j	$d_{ak} = \min\begin{bmatrix}27,30\end{bmatrix} = 27$ $d_{al} = \min\begin{bmatrix}\infty,30\end{bmatrix} = 30$	k	$q_k = i$
11	k	$d_{al} = \min\begin{bmatrix}30,30\end{bmatrix} = 30$	l	$q_l = j$

1.4. Method for finding the shortest path between all pairs of nodes

When solving practical transportation problems, it is often necessary to calculate the shortest path between all pairs of nodes in a transportation network. This problem can also be solved using the algorithm for finding the shortest path between one node and all other nodes in the transportation network. If the network han n nodes, then the Dijkstra algorithm must be applied n times, taking a different node each time as the starting node. Then after going through the algorithm n times, the shortest path will be calculated between all pairs of nodes in the network. However, the Dijkstra algorithm is rarely used to determine the shortest path between all pairs of nodes since a special algorithm exists for this type of problem. This is the algorithm developed by R.W. Flyod in 1962.[20]

Let our problem be to find the shortest path between all nodes in transportation network $G(N,A)$. We denote all nodes of the network by positive whole numbers 1, 2, ..., n. We now introduce D_o which is the beginning matrix of the shortest path lengths and Q_o which is the predecessor matrix.

We denote by d_{ij}^k the length of the shortest path from node i to node j which is found in the k-th passage through the algorithm, and by q_{ij}^k the immediate predecessor node of node j on the shortest path from node i which is also discovered on the k-th passage. Elements d_{ij}^o of matrix D_o are defined in the following manner :

If a branch exists between node i and node j,

the length of the shortest path between these nodes equals length l(i,j) of branch (i,j) which connects them. Should there be several branches between nodes i and node j, the length of the shortest path d_{ij}^o must equal the length of the shortest branch, i.e.:

$$d_{ij}^o = \min \left[l_1(i,j), l_2(i,j), \ldots, l_m(i,j) \right]$$

where m is the number of branches between node i and node j.

It is clear that $d_{ij}^o = 0$ when $i = j$. In the case when there is no branch between node i and node j, we have no information at the beginning concerning the length of the shortest path between these two nodes so we treat them as though they were infinitely far from each other, i.e. that the following is true for such pairs of nodes :

$$d_{ij}^o = \infty$$

Elements q_{oj}^o of the predecessor matrix Q_o are defined as follows :

First, we assume that $q_{ij}^o = i$, for $i \neq j$, i.e. that for every pair of nodes (i,j) for $i \neq j$, the immediate predecessor of node j on the shortest path leading from node i to node j is actually node i.

Since we have defined the elements of matrixes D_o and Q_o, we can now take a look at Flyod's algorithm, which contains the following steps :

Step 1 : Let k = 1

Step 2 : We calculate elements d_{ij}^k of the shortest path length matrix found after the k-th passage through algorithm D_k using the following equation :

$$d_{ij}^k = \min\left[d_{ij}^{k-1}, \; d_{ik}^{k-1} + d_{kj}^{k-1}\right]$$

Step 3 : Elements q_{ij}^k of predecessor matrix Q_k found after the k-th passage through the algorithm are calculated as follows :

$$q_{ij}^k = \begin{cases} q_{kj}^{k-1}, & \text{for } d_{ij}^k \neq d_{ij}^{k-1} \\ q_{ij}^{k-1}, & \text{otherwise} \end{cases}$$

Step 4 : If $k = n$, the algorithm is finished. If $k < n$, increase k by 1, i.e. $k = k + 1$ and return to Step 2.

Let us now look at the algorithm in a little more detail. In Step 2, each time we go through the algorithm we are checking as to whether a shorter path exists between nodes i and j other than the path we already know about which was established during one of the earlier passages through the algorithm. If we establish that $d_{ij}^k \neq d_{ij}^{k-1}$, i.e. if we establish during the k-th passage through the algorithm that the length of the shortest path d_{ij}^k between nodes i and j is less than the length of the shortest path d_{ij}^{k-1} known previous to the k-th passage, we have to change the immediate predecessor node to node j. Since the length of the new shortest path is :

$$d_{ij}^k = d_{ik}^{k-1} + d_{k,j}^{k-1}$$

it is clear that in this case node k is the new immediate predecessor node to j, and therefore :

$$q_{ij}^k = q_{kj}^{k-1}$$

24 TRANSPORTATION NETWORKS

This is actually done in the third algorithmic step. It is also clear that the immediate predecessor node to node j does not change if, at the end of Step 2, we have established that no other new, shorter path exists. This means that :

$$q_{ij}^{k} = q_{ij}^{k-1} \quad \text{for} \quad d_{ij}^{k} = d_{ij}^{k-1}$$

When we go through the algorithm n times (n is the number of nodes in the transportation network), elements d_{ij}^{n} of final matrix D_n will constitute the shortest path's lengths between pairs of nodes (i,j) and elements q_{in}^{n} of matrix Q_n will enable us to determine all of the nodes which are on the shortest path going from node i to node j.

E x a m p l e : Determine the shortest paths between all pairs of nodes on transportation network T(N,A) shown in Figure 19.

Branch lengths are shown on the figure.

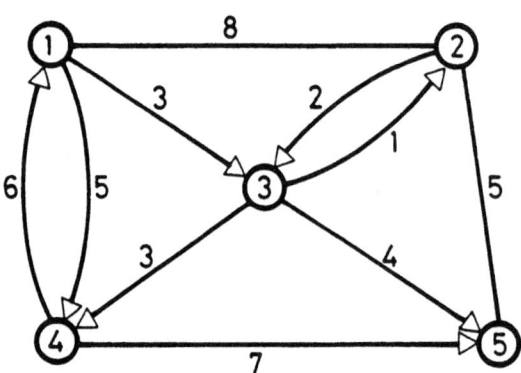

FIGURE 19. Transportation network T(N,A).

SHORTEST PATHS IN TRANSPORTATION NETWORKS 25

Starting matrix D_o is as follows :

$$D_o = \begin{array}{c} \\ 1 \\ 2 \\ 3 \\ 4 \\ 5 \end{array} \begin{array}{c} \begin{array}{ccccc} 1 & 2 & 3 & 4 & 5 \end{array} \\ \left[\begin{array}{ccccc} 0 & 8 & 3 & 5 & \infty \\ 8 & 0 & 2 & \infty & 5 \\ \infty & 1 & 0 & 3 & 4 \\ 6 & \infty & \infty & 0 & 7 \\ \infty & 5 & \infty & \infty & 0 \end{array} \right] \end{array}$$

All elements along the main diagonal of matrix D_o equal zero since by definition $d^o_{ij} = 0$ for $i = j$. We note element $d^o_{1,2}$ of matrix D_o. This element equals 8 since the length of the branch connecting nodes 1 and 2 is 8. Element $d^o_{3,1}$ equals infinity since the network has no branch which is oriented from node 3 to node 1, but rather the orientation is from node 1 to node 3. Element $d^o_{5,1}$ of matrix D_o equals infinity as well since there is no branch linking nodes 5 and 1.

Starting matrix Q_o is as follows :

$$Q_o = \begin{array}{c} \\ 1 \\ 2 \\ 3 \\ 4 \\ 5 \end{array} \begin{array}{c} \begin{array}{ccccc} 1 & 2 & 3 & 4 & 5 \end{array} \\ \left[\begin{array}{ccccc} - & 1 & 1 & 1 & 1 \\ 2 & - & 2 & 2 & 2 \\ 3 & 3 & - & 3 & 3 \\ 4 & 4 & 4 & - & 4 \\ 5 & 5 & 5 & 5 & - \end{array} \right] \end{array}$$

First we note that node i is the immediate predecessor of node j on the shortest path leading from node i to node j (for $i \neq j$). For this reason we have, for example :

26 TRANSPORTATION NETWORKS

$$q^o_{2,1} = q^o_{2,3} = q^o_{2,4} = q^o_{2,5} = 2$$

We now go to the first algorithmic step. Let k = 1. We calculate the elements of the first three rows of matrix D_1. Going through the algorithm the first time, we have :

$$d^1_{1,2} = \min\left[d^o_{1,2};\ d^o_{1,1} + d^o_{1,2}\right] = \min\left[8;\ 0 + 8\right] = 8$$

$$d^1_{1,3} = \min\left[d^o_{1,3};\ d^o_{1,1} + d^o_{1,3}\right] = \min\left[3;\ 0 + 8\right] = 3$$

$$d^1_{1,4} = \min\left[d^o_{1,4};\ d^o_{1,1} + d^o_{1,4}\right] = \min\left[5;\ 0 + 5\right] = 5$$

$$d^1_{1,5} = \min\left[d^o_{1,5};\ d^o_{1,1} + d^o_{1,5}\right] = \min\left[\infty;\ 0 + \infty\right] = \infty$$

$$d^1_{2,1} = \min\left[d^o_{2,1};\ d^o_{2,1} + d^o_{1,1}\right] = \min\left[8;\ 8 + 0\right] = 8$$

$$d^1_{2,3} = \min\left[d^o_{2,3};\ d^o_{2,1} + d^o_{1,3}\right] = \min\left[2;\ 8 + 3\right] = 2$$

$$d^1_{2,4} = \min\left[d^o_{2,4};\ d^o_{2,1} + d^o_{1,4}\right] = \min\left[\infty;\ 8 + 5\right] = 13$$

$$d^1_{2,5} = \min\left[d^o_{2,5};\ d^o_{2,1} + d^o_{1,5}\right] = \min\left[5;\ 8 + \infty\right] = 5$$

$$d^1_{3,1} = \min\left[d^o_{3,1};\ d^o_{3,1} + d^o_{1,1}\right] = \min\left[\infty;\ \infty + \infty\right] = \infty$$

$$d^1_{3,2} = \min\left[d^o_{3,2};\ d^o_{3,1} + d^o_{1,2}\right] = \min\left[1;\ \infty + 8\right] = 1$$

$$d^1_{3,4} = \min\left[d^o_{3,4};\ d^o_{3,1} + d^o_{1,4}\right] = \min\left[3;\ \infty + 5\right] = 3$$

$$d^1_{3,5} = \min\left[d^o_{3,5};\ d^o_{3,1} + d^o_{1,5}\right] = \min\left[4;\ \infty + \infty\right] = 4$$

SHORTEST PATHS IN TRANSPORTATION NETWORKS 27

Matrix D_1 is as follows :

$$D_1 = \begin{bmatrix} & 1 & 2 & 3 & 4 & 5 \\ 1 & 0 & 8 & 3 & 5 & \infty \\ 2 & 8 & 0 & 2 & ⑬ & 5 \\ 3 & \infty & 1 & 0 & 3 & 4 \\ 4 & 6 & ⑭ & ⑨ & 0 & 7 \\ 5 & \infty & 5 & \infty & \infty & 0 \end{bmatrix}$$

Matrix elements which changed values compared to the values they had in matrix D_0 are circled.

So, for example, the shortest distance between nodes 2 and 4 is 13 after the first algorithic step. In starting matrix D_0 this distance was ∞. Since

$$d^1_{2,4} = d^0_{2,1} + d^0_{1,4} = 13 < \infty = d^0_{2,4}$$

then node 1 is the new immediate predecessor of node 4 on the shortest path from node 2 to node 4. After passing through the algorithm the first time, Q_1 looks like this :

$$Q_1 = \begin{bmatrix} & 1 & 2 & 3 & 4 & 5 \\ 1 & - & 1 & 1 & 1 & 1 \\ 2 & 2 & - & 2 & 1 & 2 \\ 3 & 3 & 3 & - & 3 & 3 \\ 4 & 4 & 1 & 1 & - & 4 \\ 5 & 5 & 5 & 5 & 5 & - \end{bmatrix}$$

After the second, third, fourth and fifth passages through the algorithm, matrices D_2, Q_2, D_3, Q_3, D_4, Q_4 and D_5, Q_5 are as follows :

28 TRANSPORTATION NETWORKS

$$D_2 = \begin{matrix} & 1 & 2 & 3 & 4 & 5 \\ 1 \\ 2 \\ 3 \\ 4 \\ 5 \end{matrix} \begin{bmatrix} 0 & 8 & 3 & 5 & \boxed{13} \\ 8 & 0 & 2 & 13 & 5 \\ \boxed{9} & 1 & 0 & 3 & 4 \\ 6 & 14 & 9 & 0 & 7 \\ \boxed{13} & 5 & \boxed{7} & \boxed{18} & 0 \end{bmatrix} \qquad Q_2 = \begin{matrix} & 1 & 2 & 3 & 4 & 5 \\ 1 \\ 2 \\ 3 \\ 4 \\ 5 \end{matrix} \begin{bmatrix} - & 1 & 1 & 1 & 2 \\ 2 & - & 2 & 1 & 2 \\ 2 & 3 & - & 3 & 3 \\ 4 & 1 & 1 & - & 4 \\ 2 & 5 & 2 & 1 & - \end{bmatrix}$$

$$D_3 = \begin{matrix} & 1 & 2 & 3 & 4 & 5 \\ 1 \\ 2 \\ 3 \\ 4 \\ 5 \end{matrix} \begin{bmatrix} 0 & \boxed{4} & 3 & 5 & \boxed{7} \\ 8 & 0 & 2 & \boxed{5} & 5 \\ 9 & 1 & 0 & 3 & 4 \\ 6 & \boxed{10} & 9 & 0 & 7 \\ 13 & 5 & 7 & \boxed{10} & 0 \end{bmatrix} \qquad Q_3 = \begin{matrix} & 1 & 2 & 3 & 4 & 5 \\ 1 \\ 2 \\ 3 \\ 4 \\ 5 \end{matrix} \begin{bmatrix} - & 3 & 1 & 1 & 3 \\ 2 & - & 2 & 3 & 2 \\ 2 & 3 & - & 3 & 3 \\ 4 & 3 & 1 & - & 4 \\ 2 & 5 & 2 & 3 & - \end{bmatrix}$$

$$D_4 = \begin{matrix} & 1 & 2 & 3 & 4 & 5 \\ 1 \\ 2 \\ 3 \\ 4 \\ 5 \end{matrix} \begin{bmatrix} 0 & 4 & 3 & 5 & 7 \\ 8 & 0 & 2 & 5 & 5 \\ 9 & 1 & 0 & 3 & 4 \\ 6 & 10 & 9 & 0 & 7 \\ 13 & 5 & 7 & 10 & 0 \end{bmatrix} \qquad Q_4 = \begin{matrix} & 1 & 2 & 3 & 4 & 5 \\ 1 \\ 2 \\ 3 \\ 4 \\ 5 \end{matrix} \begin{bmatrix} - & 3 & 1 & 1 & 3 \\ 2 & - & 2 & 3 & 2 \\ 2 & 3 & - & 3 & 3 \\ 4 & 3 & 1 & - & 4 \\ 2 & 5 & 2 & 3 & - \end{bmatrix}$$

$$D_5 = \begin{matrix} & 1 & 2 & 3 & 4 & 5 \\ 1 \\ 2 \\ 3 \\ 4 \\ 5 \end{matrix} \begin{bmatrix} 0 & 4 & 3 & 5 & 7 \\ 8 & 0 & 2 & 5 & 5 \\ 9 & 1 & 0 & 3 & 4 \\ 6 & 10 & 9 & 0 & 7 \\ 13 & 5 & 7 & 10 & 0 \end{bmatrix} \qquad Q_5 = \begin{matrix} & 1 & 2 & 3 & 4 & 5 \\ 1 \\ 2 \\ 3 \\ 4 \\ 5 \end{matrix} \begin{bmatrix} - & 3 & 1 & 1 & 3 \\ 2 & - & 2 & 3 & 2 \\ 2 & 3 & - & 3 & 3 \\ 4 & 3 & 1 & - & 4 \\ 2 & 5 & 2 & 3 & - \end{bmatrix}$$

Matrices D_5 and Q_5 furnish us with complete information on the lengths of the shortest paths and the nodes on those paths between all pairs of nodes in the transportation network. For example, the shortest path from node 5 to node 4 has a length of 10. Node 3 is the immediate predecessor of node 4 on this path since $q_{5,4}^5 = 3$.

Node 2 is the immediate predecessor of node 3 on the shortest path from node 5 to node 3 since $q_{5,3}^5 = 2$. And since $q_{5,2}^5 = 5$, the shortest path from node 5 to node 4 is $(5, 2, 3, 4)$.

1.5. Method for determining shortest paths when the transportation network contains two types of branches

Two types of branches sometimes appear in certain transportation networks. In addition to ordinary branches, such networks also have special "cheap" branches. On these branches, transportation costs can be cheaper, travel time can be significantly shorter than on normal branches of the same length, etc. For example, if, when transporting freight, both trucks and trains are used between two nodes, then the train network can be considered the special cheap branch since railway transportation is less expensive. In this case, the road network represents ordinary branches. Transfers from cheap to ordinary or from ordinary to cheap networks take place in "stations". All pairs of stations do not have to be linked to each other by cheap branches and a special cheap network can consist of several unconnected parts. It is clear that we are also interested in the shortest paths between all pairs of nodes for this type of transportation network as well. The algorithm for solving this problem was given by D. Blumenfeld and M. Landau in 1972.[7]

First we find the cheapest (shortest) path going from some node K to some node L which contains two types of branches. We denote respectively by D(K,L), A(K,I) and B(J,L) minimum travel costs between pairs of nodes (K,L), (K,I) and (J,L) when using only ordinary branches for traveling. We also denote by C(I,J) minimum travel costs between node I and node J along the cheap path which is entered at node I and exited at node J. We assume that travel costs are known for all pairs of nodes in the network and for all "stations". Travel costs from node K to node L are :

    D(K,L) if only ordinary branches are used when traveling from node K to node L, or

    A(K,I) + C(I,J) + B(J,L) if the cheap path is entered at node I and exited at node J when traveling from node K to node L.

The cheapest (shortest) path from node K to node L can be found by calculating and comparing travel costs from K to L for all possible combinations of "stations" I and J. However, this entails a large number of calculations and comparisons, even when there are only a few special cheap branches.

We denote by N (K,J) the "station" where the cheapest path from K to J first enters the cheap branches. If we only use ordinary branches on the cheapest path from K to J, then N(K,J) = J. We also note the path from node I to node L and designate by $M(I,L)$ the station where the cheapest path from I to L leaves the network of cheap paths. If the cheapest path from I to L uses only ordinary branches, it is clear that $M(I,L) = I$.

From the set of all possible paths from node K to node L we single out subset S of paths from K to L which contains the following paths :
  (1) the cheapest path containing only ordinary branches,
  (2) the path which enters the network of cheap paths at node I, leaves the cheap network at node J and which satisfies the following conditions :
    (I) $I \neq J$
    (II) $M(K,J) = I$
    (III) $M(I,L) = J$

The cheapest path from node K to node L must belong to subset S. If we only use ordinary branches when going from K to L, then this path belongs to subset S (condition (1)). Now we note the situation when special cheap branches enter into the composition of the cheapest path from node K to node L. If in this case condition (2-I) is not satisfied, i.e. if $I = J$, then this path enters and exits the cheap network at this station. The path through node I which does not use the cheap network will then be cheaper by $C(I,I)$ (or the price will be the same if $C(I,I) = 0$). So, if the cheapest path from K to L uses the cheap network, then it is essential that $I \neq J$. We assume that the cheapest path from K to L uses the cheap network,

32 TRANSPORTATION NETWORKS

that condition (2-I) is satisfied and that condition (2-II) is not satisfied, i.e. we assume that path K→I→J→L first enters the cheap network at some node I*, and not at node I. We have therefore assumed that :

$N(K,J) = I^* \neq I$

In this case, path K→I*→J will be cheaper than path K→I→J and path K→I*→J→L will be cheaper than path K→I→J→L. We conclude, then, that if the cheapest path from K to L uses the cheap network, entering it at node I and exiting it at node J, then $N(K,J) = I$ has to be true. In the same manner, we can prove that condition (2-III) must also be satisfied, i.e. $M(I,L) = J$.

E x a m p l e : Using the Blumenfeld and Landau algorithm, find the cheapest path from node K to node L in the transportation network composed of ordinary branches and special branches shown on Fig. 20. Solid lines denote cheap branches and dotted lines denote ordinary branches. Nodes K and L are also connected to all stations by ordinary branches.

The following costs are known :

$$D(K,L) = 14$$

A (K,1)	= 2		B (1,L)	= 14
A (K,2)	= 5		B (2,L)	= 12
A (K,3)	= 8		B (3,L)	= 9
A (K,4)	= 10		B (4,L)	= 6
A (K,5)	= 8		B (5,L)	= 6
A (K,6)	= 11		B (6,L)	= 3

SHORTEST PATHS IN TRANSPORTATION NETWORKS   33

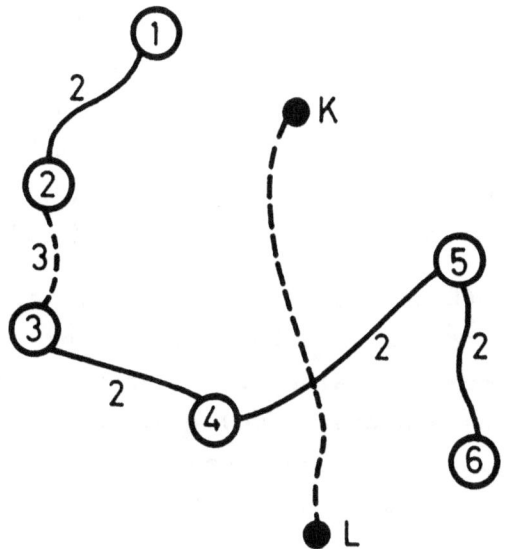

FIGURE 20. A transportation network composed
of ordinary branches and special cheap branches.

Travel costs $C(I,J)$ between nodes I and J in
the network of cheap branches are additive. For
example, $C(2,5) = C(2,3) + C(3,4) + C(4,5) =$
$3 + 2 + 2 = 7$.

We now define stations $N(K,J)$ where the cheapest
path from node K to node J first enters the network
of cheap paths. So we have, for example :

   $N(K,1) = 1$, $N(K,2) = 1$, $N(K,3) = 1$,
         $N(K,4) = 1$ $N(K,5) = 5$

since it is less expensive from node K to enter the
network of cheap paths at node 5 compared to nodes
1, 2, 3, or 4.

We also have $N(K,6) = 5$. Now we define stations

M(I,L) in the same manner. For our network, stations N(K,J) and M(I,L) are as follows:

N (K,1)	=	1	M (1,L)	=	4
N (K,2)	=	1	M (2,L)	=	4
N (K,3)	=	1	M (3,L)	=	4
N (K,4)	=	1	M (4,L)	=	4
N (K,5)	=	5	M (5,L)	=	6
N (K,6)	=	5	M (6,L)	=	6

Since $N(K,4) = 1$ and $M(1,L) = 4$, path K→1→4→L enters into the composition of subset S. Path K→5→6→L also becomes part of subset S since $N(K,6) = 5$ and $M(5,L) = 6$. Path K→L which does not go through the cheap network also enters the subset. Costs along these paths are given in Table II.

TABLE II. Paths in subset S and their costs

path	costs
K → L	14
K→1→4→L	15
K→5→6→L	13

As can be seen, the cheapest path from node K to node L is K→5→6→L which contains both ordinary and cheap branches.

1.6. Finding the shortest path in a probabilistic network

Branch length, as one of the most important parameters in the transportation network, can be a random variable. This is particularly true for those cases when branch length, which as we mentioned earlier is a general concept, stands for travel time. In many problems, travel time along a certain branch is a random variable and a probabilistic approach is required to solve these problems. A network in which individual branch lengths are random variables with a certain probability distribution are called probabilistic networks. Solving problems on a probabilistic network is much harder and more complex than on a deterministic network, primarily due to the difficult calculations which can arise. One of the main reasons for this is that calculating the shortest path between two points in a probabilistic network is extremely complex since the shortest path algorithm is also used to solve many different problems on the network.

Let us note nonoriented network $G(N,A)$ shown on Fig. 21.

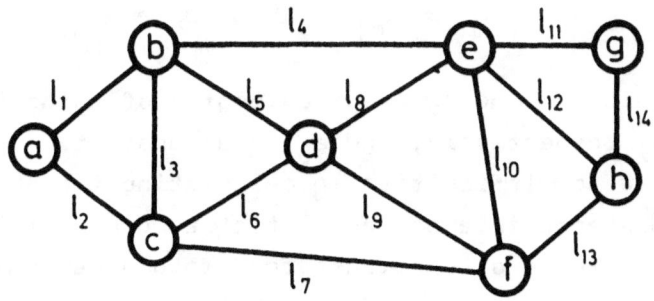

FIGURE 21. Nonoriented network $G(N,A)$.

Let the lenght of branches i, L_i be the random variables which are subject to a certain probability distribution. We want to find the shortest path between specific nodes of the probabilistic network, say node $a \in N$ and node $h \in N$. Let there be k alternative paths $S_1, S_2, \ldots, S_k$ between nodes a and h and these paths can have several common branches. We denote the length of path S_j by A_j. This means that :

$$A_j = \sum_{\text{all branches} \in S_j} L_i$$

We denote the shortest path between nodes a and h by d_{ah}. Therefore :

$$d_{ah} = \min (A_1, A_2, \ldots, A_k)$$

When choosing the shortest path in a probabilistic network, individual alternative paths can be compared. To this effect, we say that path S_1, a_o is better than S_2 if :

$$P \left[A_1 \geq a_o \right] < P \left[A_2 \geq a_o \right]$$

for which A_1 and A_2 are the lengths of paths S_1 and S_2 respectively, and a_o is a constant.

Due to difficulties in calculating the shortest distance in a probabilistic network, this type of network is often transformed into a deterministic network to the effect that it operates with branch length means. Also, in some cases branch length is treated as a discrete random variable which can take on only a certain number value.[29] By p_a we denote the probability that the length of branch (i,j), L(i,j) equals l_a, i.e. :

where
$$\Pi\left[L(i,j) = l_a\right] = P_a \quad \text{for } a = 1,2,\ldots,b$$
$$\sum_{a=1}^{b} P_a = 1$$

In this case, branch length can only take on one of the values $l_1, l_2, \ldots, l_b$.

A direct result of the assumption that branch lengths can take on only a certain number value is that the network can be put into a final number of states. Let m be the number of possible states in which a network can be found. One state differs from the other if at least one branch in the network has a different length. The possible state of networks is denoted by $R_1, R_2, \ldots, R_m$. There is a probability of P_c, $c = 1,2,\ldots,m$ that the network is in some state R_c.

We denote the length of branch (i,j) by $l_c(1,j)$ when the network is in state R_c, and the shortest distance between points p and q in this state is denoted by $d_c(p,q)$.

The network which is now in some state R_c can be treated as a deterministic network, and by knowing lengths $l_c(i,j)$ of branches (i,j) and applying the appropriate algorithm for finding the shortest path, we can calculate the length of shortest path $d_c(p,q)$ between any two points p and q in the network.

The mean length of the shortest path between any two points $p,q \in G$ is :

$$M\left[D(p,q)\right] = \sum_{c=1}^{m} P_c \, d_c(p,q)$$

38 TRANSPORTATION NETWORKS

Difficulties in calculating primarily depend on the total number of states m in which the probabilistic network can be put.

E x a m p l e : Let us consider network M(N,A) in Fig. 22.

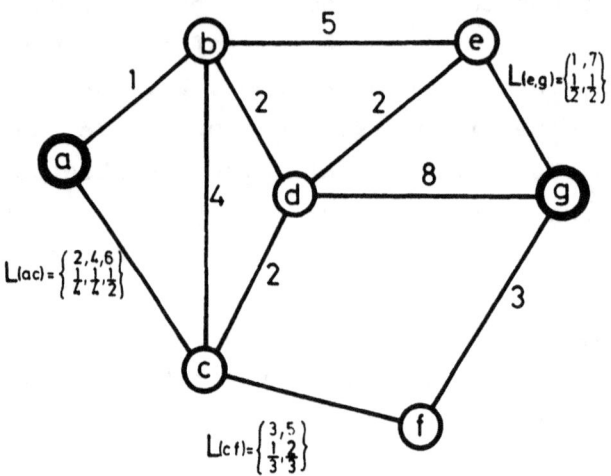

FIGURE 22. Probabilistic network M(N,A).

The lengths of branches (a,c), (c,f) and (e,g) are random variables with certain probability distributions which are noted on the Figure next to these branches. For example, the probability that the length of branch (c,f) is 3 is :

$$P\left[L(c,f) = 3\right] = \frac{1}{3}$$

We assume that the value taken on by a branch

does not depend on the values taken on by other branches in the network and we calculate shortest path $D(a,g)$ between node a and node g.

The probabilistic network under study can be put into 12 possible states. Table III shows these 12 states, branch lengths in each of them, the corresponding probability that the network is in any particular state, the shortest paths and their lengths in each state.

TABLE III States, branch lengths and shortest paths in network $M(N,A)$

State c	Branch length (a,c)	(c,f)	(e,g)	p_c	Shortest path	$d_c(a,g)$
1	2	3	1	$\frac{1}{24}$	(a,b,d,e,g)	6
2	2	3	7	$\frac{1}{24}$	(a,c,f,g)	8
3	2	5	1	$\frac{1}{12}$	(a,b,d,e,g)	6
4	2	5	7	$\frac{1}{12}$	(a,c,f,g)	10
5	4	3	1	$\frac{1}{24}$	(a,b,d,e,g)	6
6	4	3	7	$\frac{1}{24}$	(a,c,f,g)	10
7	4	5	1	$\frac{1}{12}$	(a,b,d,e,g)	6
8	4	5	7	$\frac{1}{12}$	(a,b,d,g)	11
9	6	3	1	$\frac{1}{12}$	(a,b,d,e,g)	6
10	6	3	7	$\frac{1}{12}$	(a,b,d,g)	11
11	6	5	1	$\frac{1}{6}$	(a,b,d,e,g)	6
12	6	5	7	$\frac{1}{6}$	(a,b,d,g)	11

40 TRANSPORTATION NETWORKS

The mean length of the shortest path between node a and node g is

$$M\left[D(a,g)\right] = \sum_{c=1}^{i_2} P_c \cdot d_c (a,g) = 8,25$$

1.7. Method for finding minimum spanning trees

As we have already mentioned, a spanning tree of graph $G(N,A)$ means any tree of graph G which contains the entire set of nodes N. Fig. 23 shows transportation network $L(N,A)$ and one of its spanning trees.

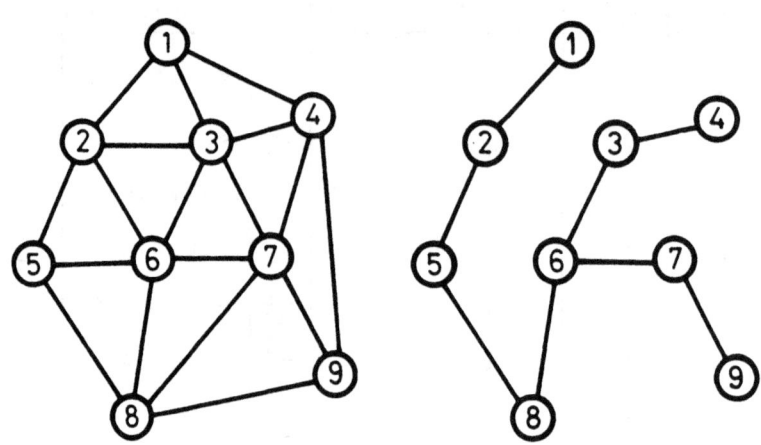

FIGURE 23. Transportation network $L(N,A)$ and one of its spanning trees.

The problem of the minimum spanning tree is to find out of all possible spanning trees of graph $G(N,A)$ the one which connects all nodes with the shortest total lengths of branches.

SHORTEST PATHS IN TRANSPORTATION NETWORKS 41

The minimum spanning tree problem can be directly applied to designing transportation networks. It also plays an important part in solving more complex problems, such as the problem of covering nodes or rotating transportation vehicles on a network.

The following example illustrates the importance of determining the minimum spanning tree when designing transportation networks. Let there be n cities which must be connected by roads. In this case the minimum spanning tree can be interpreted as the minimum road length needed which will directly or indirectly connect all n cities, assuming that all roads start and finish in a pair of cities.

There are several very efficient algorithms which successfully solve the problem of finding the minimum spanning tree. All of these algorithms are founded on the fact that in nonoriented network G the shortest branch leaving any node must also belong to the minimum spanning tree. Interested readers will find proof of this in reference (29).

The algorithm for finding the minimum spanning tree is comprised of the following algorithmic steps :

Step 1 : Construction of the minimum spanning tree starts in arbitrary node i. Then the closest node to i is found. Let it be some node j. Node j and branch (i,j) are included in the minimum spanning tree. Then all branches are removed (if they exist) which make creating the tree

impossible (since they would form
cycles).

Step 2 : If all nodes are connected, the minimum spanning tree has been found. If any isolated nodes still exist, we go on to step 3.

Step 3 : An isolated node which is closest to the thus-formed minimum spanning tree is included along with its branch into the composition of the minimum spanning tree. Should branches exist which hinder the creation of the tree, they should be broken off. Upon completion of this step, we return to Step 2.

E x a m p l e : Determine the minimum spanning tree of transportation network $B(N,A)$ shown in Fig. 24 (individual branch lengths are noted).

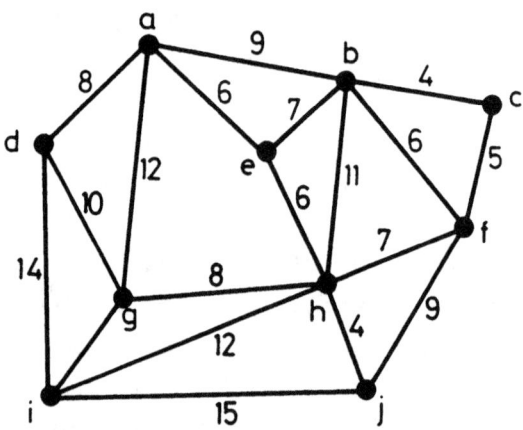

FIGURE 24 Transportation network $B(N,A)$.

We start construction of the shortest spanning tree at node a. Since node e is the closest to node a (the distance between them is 6), the minimum spanning tree now includes nodes a and e and branch (a,e) as shown on Fig. 25.

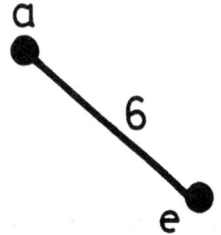

FIGURE 25. Transportation network B(N,A) after first passage through the algorithm.

Since there are still many isolated nodes, we go on to step 3. We choose the node which is closest to the tree that has been formed thus far and include it along with its corresponding branch. Nodes b, d and g are isolated nodes connected to node a. Node d is the closest of these to node a. Isolated nodes b and h are connected to node e. Node h is closer to node e than node b. Since node h is closer to node e than node d is to node a, we include branch (e,h) and node h into the composition of the minimum spanning tree (Fig. 26).

We continue by including the next closest node which is node j (Fig. 27).

We can now include either node b or node f (both are at an equal distance of 7 from the closest node already included in the tree). We choose node b (it is the first in alphabetical order) (Fig. 28). If we included node f into the

44 TRANSPORTATION NETWORKS

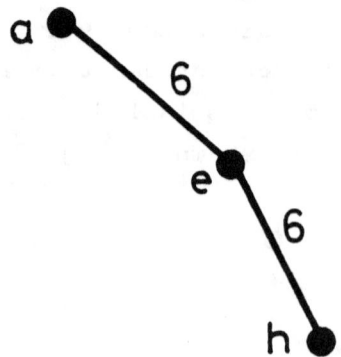

FIGURE 26. Transportation network $B(N,A)$ after second passage through the algorithm.

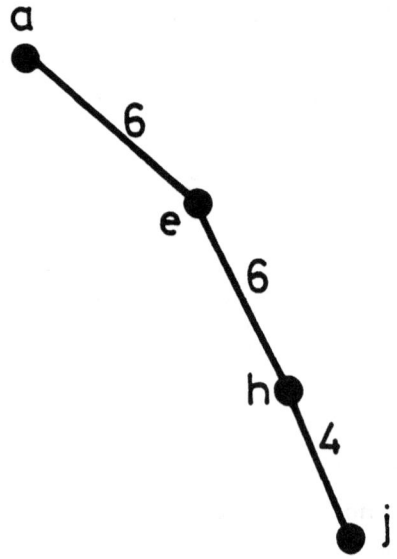

FIGURE 27. Transportation network $B(N,A)$ after third passage through the algorithm.

tree, we would get a different tree in the end
from the one we will get with node b. However,
both these trees have the same total length of
all branches.

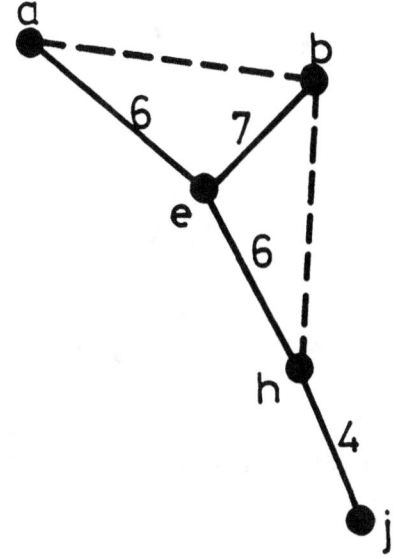

FIGURE 28.   Transportation network B(N,A)
after the fourth passage through the algorithm.

Branches (a,b) and (b,h) would hinder the
creation of a tree (since they would form a
cycle) so we break them off and remove them.

Node c is the next to enter the tree (Fig. 29).
Then node f and branch (c,f) are included into the
tree thereby removing branches (b,f), (f,h) and
(f,j) which would hinder the creation of a tree
(Fig. 30).

46 TRANSPORTATION NETWORKS

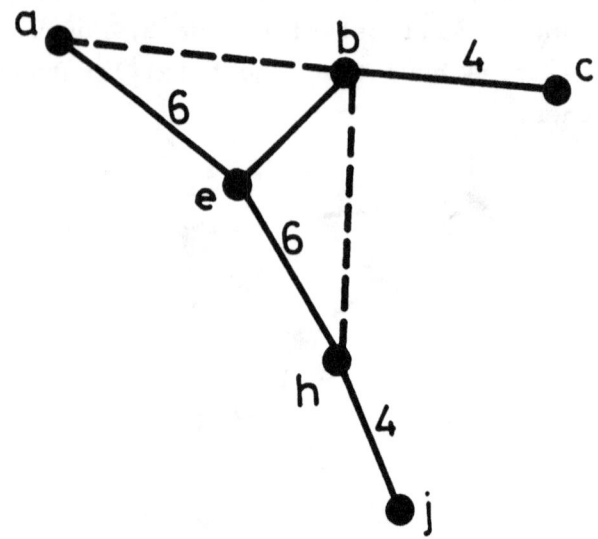

FIGURE 29. Transportation network B(N,A) after the fifth passage through the algorithm.

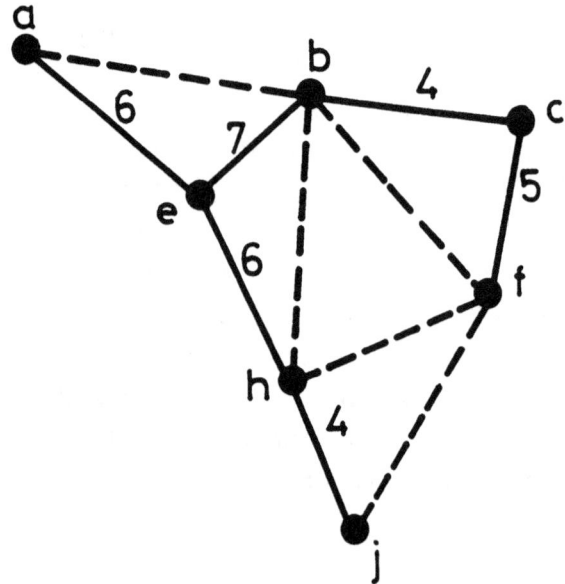

FIGURE 30. Transportation network B(N,A) after the sixth passage through the algorithm.

SHORTEST PATHS IN TRANSPORTATION NETWORKS 47

Node d and branch (a,d) are now included
into the tree (Fig. 31)

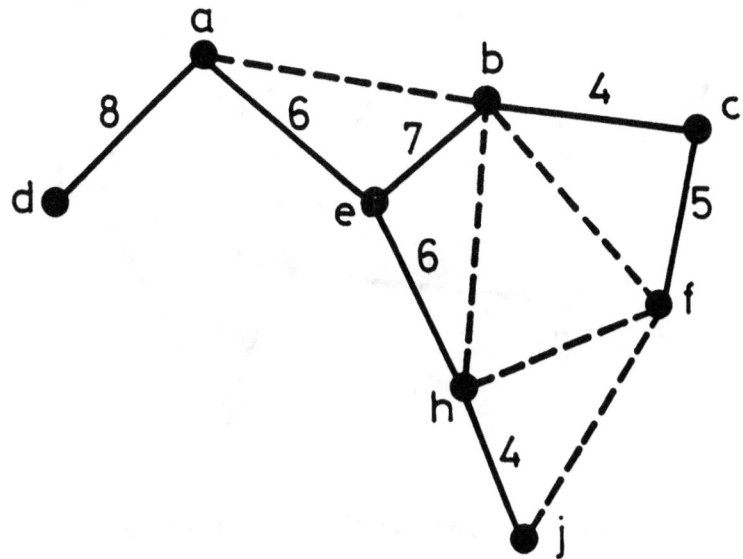

FIGURE 31.   B(N,A) seventh time through.
By including node g and branch (g,h) and removing
branches (d,g) and (a,g) Fig. 32 is formed.

And finally, by including node i and branch
(g,i) and removing branches (d,i), (h,i) and
(i,j) we get the minimum spanning tree (Fig. 33).

48 TRANSPORTATION NETWORKS

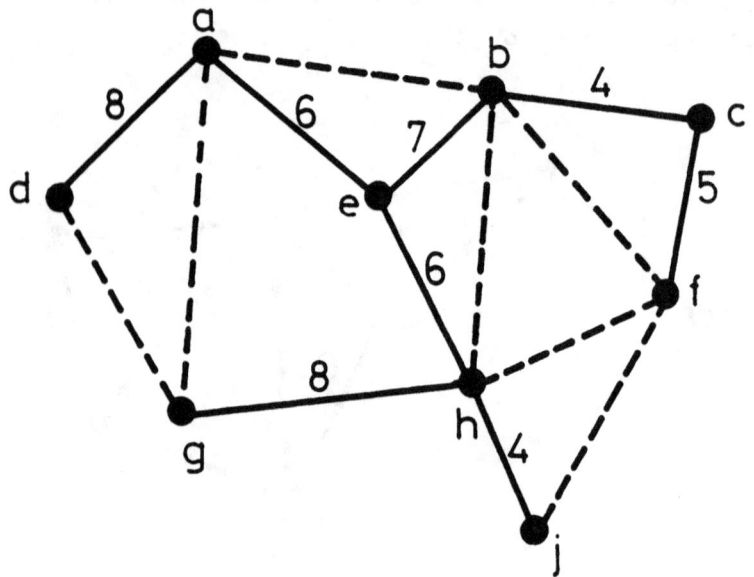

FIGURE 32. B(N,A) eighth time through.

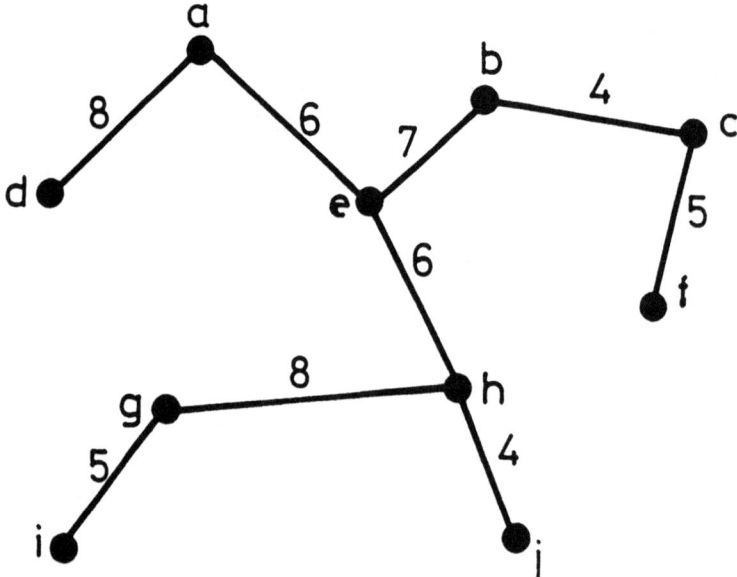

FIGURE 33. B(N,A) minimum spanning tree.

2 Transportation Network Flows

2.1. Flows on transportation networks

Transportation flows indicate quantities of goods or numbers of passengers passing through some point during a specific unit of time, i.e. :

$$f_{ij}(t) = \frac{\left[\begin{array}{l}\text{quantity passing through a point on}\\ \text{branch } (i,j) \text{ from time } t \text{ to } d + dt\end{array}\right]}{dt}$$

or :

$$f_{ij}(t) = \frac{d\left[\begin{array}{l}\text{cumulative amount passing through}\\ \text{point on branch } (i,j) \text{ from } 0 \text{ to } t\end{array}\right]}{dt}$$

Flows are values which change over time. Changes are notable by month, week, day and finally by hour. Figure 34 shows a flow as a function of the time of day which is characteristic of many phenomena in traffic and transport.

On this figure, f(t) denotes the flow and T the time interval during which requests for transportation are noted.

The shape of the flow appears most often by counting or by taking surveys. Flows which are a function of the time of day are most often characterized by two maximums, or "peaks" - a morning and an evening.

A flow is said to have a steady state if the intensity on the branch under observation during

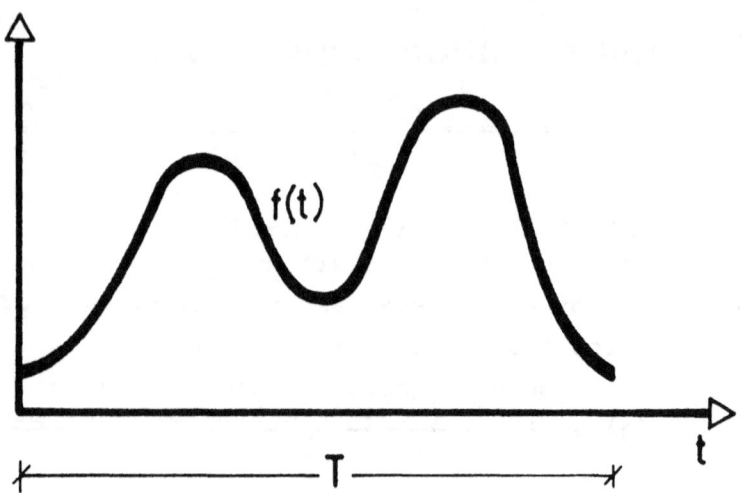

FIGURE 34. Flow as a function of the time of day.

the observation period is approximately constant. Steady state flows are primarily dealt with in the theory of transportation networks. The majority of the transportation networks studied are observed over a short period of time so that changes in intensity over time are disregarded and each branch (i,j) is characterized by a unique flow value $f_{ij} > 0$.

2.2. Conservation laws on transportation networks

When a transportation network is in a steady state, flows cannot be formed suddenly or destroyed suddenly. For example, in the case of an intermediate node, the sum of all flows entering the intermediate node equals the sum of all flows exiting it. Nodes which create, or attract traffic are normally called centroids. A closer observation of these nodes reveals that the sum of all flows exiting a centroid equals the flows which are created in the centroid, and the sum of all flows entering the centroid equals the flows which the centroid attracts.

Let us consider transportation network $G(N, L)$. We denote the flow on oriented branch (i,j) by f_{ij}. We also denote the flow created in centroid i by a_i and the flow attracted to centroid i by b_i. Quantities f_{ij}, a_i and b_i are nonnegative. We denote the set of nodes "after" node i by $A(i)$ and the set of nodes "before" node i by $B(i)$, and let :

$$A(i) = \left[j \mid j \in N, (i,j) \in L \right]$$
$$B(i) = \left[j \mid j \in N, (j,i) \in L \right]$$

This means that set $A(i)$ includes all nodes that are finishing points for branches exiting node i, and set $B(i)$ includes all nodes that are starting points for branches finishing in node i.

The flow conservation law on transportation networks can be written in the form of the following equations :

$$\left. \begin{array}{c} \sum\limits_{A(i)} f_{ij} = a_i \\ \sum\limits_{B(i)} f_{ji} = b_i \end{array} \right\} \text{if } i \text{ is a centroid}$$

$$\sum\limits_{A(i)} f_{ij} - \sum\limits_{B(i)} f_{ji} = 0 \Big\} \text{if } i \text{ is an intermediate node}$$

52 TRANSPORTATION NETWORKS

In order for the above system of equations to be solved, the sum of flows emerging from all nodes on the transportation network created by flows $\sum a_i = v$ must equal the sum of flows entering nodes on the network which attract flows $\sum b_i = v$. The number of unknows in the system of conservation equations is much larger than the number of equations, thus the system of flow conservation equations does not have a unique solution. This means that the same transportation network can be characterized by several different sets of flows on its branches.

E x a m p l e

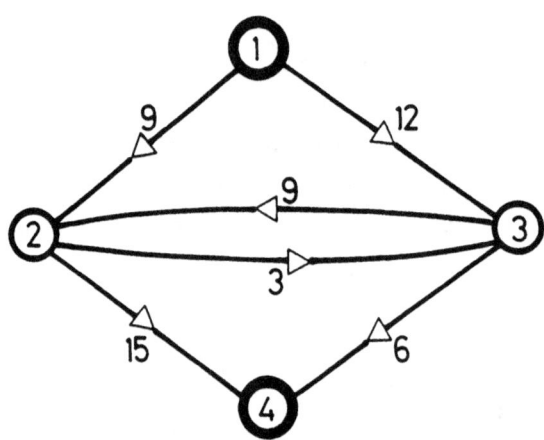

FIGURE 35. Applying conservation laws to a transportation network.

Figure 35 shows a transportation network with centroids 1 and 4 and intermediate nodes 2 and 3.

Flows along each branch are noted on the figure. For example, the conservation law for intermediate node 2 can be easily shown as follows :

$$i = 2, \quad \Lambda(2) = [3,4] \quad B(2) = [1,3]$$

$$\sum_{\Lambda(2)} f_{2j} - \sum_{B(2)} f_{j2} = f_{23} + f_{24} - f_{12} - f_{32} =$$

$$3 + 15 - 9 - 9 = 0$$

2.3. Flows on transportation networks with one source and one sink

The term transportation network source or origin signifies a node which has no branches entering it and the term transportation network sink or destination denotes a node which has no branches exiting it. A network with one source and one sink is shown on Fig. 36.

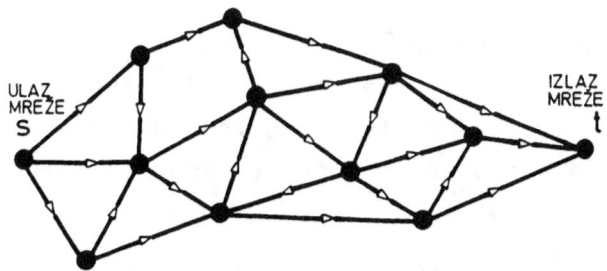

FIGURE 36. Transportation network with one source and one sink.

Mode s is the source and node t the sink of the network. Many different traffic and transportation problems can be studied within the context of a network with one source and one sink. Total flow through the transportation network

54 TRANSPORTATION NETWORKS

equals the sum of the flows of branches exiting the network's source s and entering the network's sink t. In other words :

$$\sum_{A(s)} f_{sj} = v$$

$$\sum_{A(i)} f_{ij} - \sum_{B(i)} f_{ji} = 0 \quad i \neq s, \ i \neq t.$$

$$\sum_{B(t)} f_{jb} = -v$$

where v is the total flow through the transportation network.

We now divide the set of nodes N of transportation network $G(N,A)$ into two disjunct subsets X and $\bar{X}$, i.e. :

$$X \cup \bar{X} = N \qquad X \cap \bar{X} = \emptyset$$

and let $s \in X$ and $t \in \bar{X}$.

The set of all branches $(X, \bar{X})$ starting from nodes in set X and finishing in nodes of set $\bar{X}$ is called a cut set of the transportation network. This means that :

$$[X, \bar{X}] = [(i,j) \mid (i,j) \in A, i \in X, j \in \bar{X}]$$

The flow through the transportation network cut sets is defined as follows :

$$f(X, \bar{X}) = \sum_{(i,j) \in [X, \bar{X}]} f_{ij}$$

TRANSPORTATION NETWORK FLOWS

$$f(\bar{X}, X) = \sum_{(j,i) \in [\bar{X}, X]} f_{ji}$$

Since: $v = \sum_{A(s)} f_{sj}$ and $\sum_{B(s)} f_{js} = 0$

therefore: $v = \sum_{A(s)} f_{sj} - \sum_{B(s)} f_{js}$

On the other hand:

$$\sum_{A(i)} f_{ij} - \sum_{B(i)} f_{ji} = 0, \quad i \neq s$$

$$\sum_{\substack{i \in X \\ i \neq s}} \left[\sum_{A(i)} f_{ij} - \sum_{B(i)} f_{ji} \right] = 0$$

So we can write that:

$$v = \sum_{A(s)} f_{sj} - \sum_{B(s)} f_{js} + \sum_{\substack{i \in X \\ i \neq s}} \left[\sum_{A(i)} f_{ij} - \sum_{B(i)} f_{ij} \right]$$

and

$$v = \sum_{i \in X} \sum_{A(i)} f_{ij} - \sum_{i \in X} \sum_{B(i)} f_{ji}$$

$$v = \sum_{(i,j) \in [X, \bar{X}]} f_{ij} - \sum_{(j,i) \in [\bar{X}, X]} f_{ji}$$

and finally we have

$$v = f(X, \bar{X}) - f(\bar{X}, X)$$

56 TRANSPORTATION NETWORKS

We have shown that the total flow through a transportation network equals the difference in the flows through a corresponding cut sets of the network. We would mention once more that this is satisfied only when the network source and sink belong to different subsets X and $\bar{X}$.

E x a m p l e : Let us study the transportation network on Figure 35. Node 1 is the source and node 4 is the sink. The total flow through this network is $v = 21$. We divide the nodes of this network into two disjunct subsets, i.e. let

$$X = \begin{bmatrix} 1,2 \end{bmatrix} \quad i \quad \bar{X} = \begin{bmatrix} 3,4 \end{bmatrix}$$

The network's source and sink belong to different subsets. So we have

$$f(X,\bar{X}) - f(\bar{X},X) = f_{13}+f_{23}+f_{24}-f_{32} =$$
$$12 + 3 + 15 - 9 = 21 = v$$

2.4. Branch capacities and transportation network capacities

Up to now our discussion of transportation network flows has not introduced any limitations concerning the maximum value of branch flows. However, the case must also be studied when certain limitations exist regarding the capacities of individual branches or of the entire transportation network. Flow f_{ij} of branch (i,j) must often satisfy the inequality:

$$0 \leq f_{ij} \leq c_{ij}$$

Values c_{ij} are called branch capacity.
Since :

$$f_{ij} \leq c_{ij}, \quad \forall (i,j) \in A$$

then :
$$\sum_{(i,j)\in [X,\bar{X}]} f_{ij} \leq \sum_{(i,j)\in [X,\bar{X}]} c_{ij}$$

and :
$$f(X,\bar{X}) \leq C(X,\bar{X})$$

where
$$C(X,\bar{X}) = \sum_{(i,j)\in [X,\bar{X}]} c_{ij}$$

is the capacity of the network cut set.

Therefore, the total flow through the network is always less than or equal to the capacity of the network cut set.

Since : $f(\bar{X},X) \geq 0$

then : $f(X,\bar{X}) - f(\bar{X},X) \leq C(X,\bar{X})$

and : $v \leq C(X,\bar{X})$

The entire flow through the network is always less or equal to the network cut set capacity.

2.5. Algorithm for finding the maximum flow through a transportation network

When there are limits to branch capacities on a transportation network, the problem arises of determining the maximum flow through the network. If any two nodes are connected by a branch in one direction only, the opposite prohibited direction is usually characterized by a capacity of zero. The following algorithm is most often used to determine the maximum flow between source s and sink t of a transportation network :

Step 1 : We arbitrarily determine a chain from node s to node t so that a possible flow of a certain number of passengers or quantity of freight can pass through each branch of the chain. If such a chain cannot be determined, go to Step 3. If such a chain can be determined, go to Step 2.

Step 2 : We denote the capacity of branch (i,j) in direction s→t by $c_{ij}^{(I)}$ and the capacity of branch (i,j) in direction t→s by $c_{ij}^{(II)}$. Let

$$\alpha = \min \left[c_{ij}^{(I)} \right] > 0$$

We subtract value α from all capacities $c_{ij}^{(I)}$ and add value α to all capacities $c_{ij}^{(II)}$. We replace all old capacities $c_{ij}^{(I)}$ and $c_{ij}^{(II)}$ by the new capacities obtained by subtracting or adding value α and return to Step 1.

Step 3 : Let $C = ||c_{ij}||$ be the starting capacity matrix and let $C^* = ||c^*_{ij}||$ be the last form of starting capacity C after undergoing consecutive modifications. Elements x_{ij} of the optimal flow matrix X are calculated as :

$$x_{ij} = \begin{cases} c_{ij} - c_{ij}^* & \text{for } c_{ij} > c_{ij}^* \\ 0 & \text{for } c_{ij} \leq c_{ij}^* \end{cases}$$

Maximum flow v* between source s and sink t of the transportation network is :

$$v^* = v_{max} = \sum_i x_{si} = \sum_j x_{jt}$$

E x a m p l e : determine the maximum flow between node s and node t of the transportation network shown in Fig. 37. Capacities of individual branches are shown on the figure.

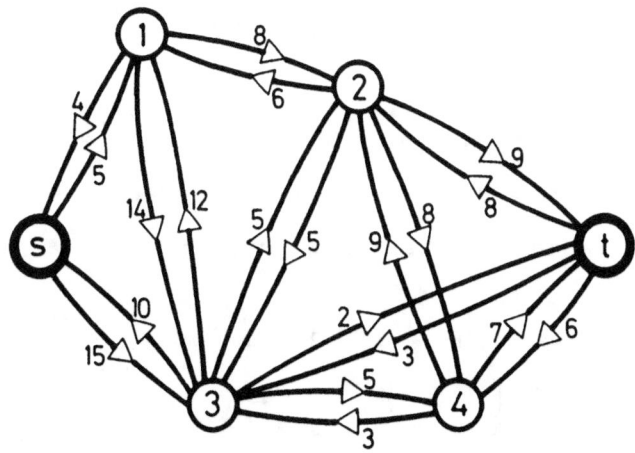

FIGURE 37. Network for which maximum flow must be found between nodes s and t.

The starting capacity matrix reads :

	s	1	2	3	4	t
s		5		15		
1	4		8	14		
2		6		5	8	9
3	10	12	5		5	2
4			9	3		7
t			8	3	6	

We arbitrarily choose chain $(s, 1, 2, t)$. Since the capacities of all branches belonging to this chain are greater than zero, we have:

60 TRANSPORTATION NETWORKS

$$\alpha = \min\left[c_{s1}^{(I)}, c_{12}^{(I)}, c_{2t}^{(I)}\right] = \min\left[5, 8, 9\right] = 5$$

New capacities are :

$c_{s1}^{(I)} = 5 - 5 = 0;\ c_{12}^{(I)} = 8 - 5 = 3;\ c_{2t}^{(I)} = 9 - 5 = 4$

$c_{s1}^{(II)} = 4 + 5 = 9;\ c_{12}^{(II)} = 6 + 5 = 11;\ c_{2t}^{(II)} = 8 + 5 = 13$

The capacity matrix now reads :

	s	1	2	3	4	t
s		0		15		
1	9		3	14		
2		11		5	8	4
3	10	12	5		5	2
4			9	3		7
t			13	3	6	

We arbitrarily choose a new chain. Let it be chain (s, 3, 4, t). In this case we have :

$$\alpha = \min\left[c_{s3}^{(I)}, c_{34}^{(I)}, c_{4t}^{(I)}\right] = \min\left[15, 5, 7\right] = 5$$

$c_{s3}^{(I)} = 15 - 5 = 10;\ c_{34}^{(I)} = 5 - 5 = 0;\ c_{4t}^{(I)} = 7 - 5 = 2$

$c_{s3}^{(II)} = 10 + 5 = 15;\ c_{34}^{(II)} = 3 + 5 = 8;\ c_{4t}^{(II)} = 6 + 5 = 11$

The capacity matrix now reads :

$$\begin{array}{c c} & \begin{array}{c c c c c c} s & 1 & 2 & 3 & 4 & t \end{array} \\ \begin{array}{c} s \\ 1 \\ 2 \\ 3 \\ 4 \\ t \end{array} & \left[\begin{array}{c c c c c c} & 0 & & 10 & & \\ 9 & & 3 & 14 & & \\ & 11 & & 5 & 8 & 4 \\ 15 & 12 & 5 & & 0 & 2 \\ & & 9 & 8 & & 2 \\ & & 13 & 3 & 11 & \end{array} \right] \end{array}$$

We now choose chain $(s, 3, 2, t)$ as the new chain.

$$\alpha = \min\left[c_{s3}^{(I)}, c_{32}^{(I)}, c_{2t}^{(I)}\right] = \min\left[10, 5, 4\right] = 4$$

$c_{s3}^{(I)} = 10 - 4 = 6;\ c_{32}^{(I)} = 5 - 4 = 1;\ c_{2t}^{(I)} = 4 - 4 = 0$

$c_{s3}^{(II)} = 15 + 4 = 19;\ c_{32}^{(II)} = 5 + 4 = 9;\ c_{2t}^{(II)} = 13 + 4 = 17$

The matrix now reads :

$$\begin{array}{c c} & \begin{array}{c c c c c c} s & 1 & 2 & 3 & 4 & t \end{array} \\ \begin{array}{c} s \\ 1 \\ 2 \\ 3 \\ 4 \\ t \end{array} & \left[\begin{array}{c c c c c c} & 0 & & 6 & & \\ 9 & & 3 & 14 & & \\ & 11 & & 9 & 8 & 0 \\ 19 & 12 & 1 & & 0 & 2 \\ & & 9 & 8 & & 2 \\ & & 17 & 3 & 11 & \end{array} \right] \end{array}$$

The next chain is $(s, 3, 7, 4, t)$.

62 TRANSPORTATION NETWORKS

$$\alpha = \min\left[c_{s3}^{(1)}, c_{32}^{(1)}, c_{24}^{(1)}, c_{4t}^{(1)}\right] = \min\left[6, 1, 8, 2\right] = 1$$

$c_{s3}^{(1)} = 6 - 1 = 5;$ $c_{32}^{(1)} = 1 - 1 = 0;$ $c_{24}^{(1)} = 8 - 1 = 7;$

$c_{s3}^{(II)} = 19 + 1 = 20;$ $c_{32}^{(II)} = 9 + 1 = 10;$ $c_{24}^{(II)} = 9 + 1 = 10;$

$c_{4t}^{(1)} = 2 - 1 = 1$ $c_{4t}^{(II)} = 11 + 1 = 12$

The new matrix is:

	s	1	2	3	4	t
s			0	5		
1	9		3	14		
2		11		10	7	0
3	20	12	0		0	2
4			10	8		1
t			17	3	12	

The new chain is $(s, 3, t)$.

$$\alpha = \min\left[c_{s3}^{(1)}, c_{3t}^{(1)}\right] = \min\left[5, 2\right] = 2$$

$c_{s3}^{(1)} = 5 - 2 = 3;$ $c_{3t}^{(1)} = 2 - 2 = 0$

$c_{s3}^{(II)} = 20 + 2 = 22;$ $c_{3t}^{(II)} = 3 + 2 = 5$

The next matrix is:

	s	1	2	3	4	t
s			0	3		
1	9		3	14		
2		11		10	7	0
3	22	12	0		0	0
4			10	8		1
t			17	5	12	

TRANSPORTATION NETWORK FLOWS 63

The next chain is $(s, 3, 1, 2, 4, t)$.

$$\alpha = \min \left[c_{s3}^{(I)}, c_{31}^{(I)}, c_{12}^{(I)}, c_{24}^{(I)}, c_{4t}^{(I)} \right] =$$

$$\min \left[3, 22, 3, 7, 1 \right] = 1$$

$c_{s3}^{(I)} = 3 - 1 = 2;\ c_{31}^{(I)} = 12-1=11;\ c_{12}^{(I)} = 3-1=2;$

$c_{s3}^{(II)} = 22+1 = 23;\ c_{31}^{(II)} = 14+1 =15;\ c_{12}^{(II)}=11+1=12;$

$c_{24}^{(I)} = 7-1 =6;\ c_{4t}^{(I)}=1-1 = 0 \quad c_{24}^{(II)}=10+1=11;\ c_{4t}^{(II)}=12+1=13$

The new modified capacity matrix is:

	s	1	2	3	4	t
s			0	2		
1	9		2	15		
2		12		10	6	0
3	23	11	0		0	0
4			11	8		0
t			17	5	13	

No other chains exist whose branches all have a capacity greater than zero, therefore we go to Step 3 of the algorithm for finding the maximum flow. Optimal flow matrix X reads:

$$X = \begin{array}{c} \\ s \\ 1 \\ 2 \\ 3 \\ 4 \\ t \end{array} \begin{array}{c} s\quad 1\quad 2\quad 3\quad 4\quad t \\ \left[\begin{array}{cccccc} & 5 & & 13 & & \\ 0 & & 6 & 0 & & \\ & 0 & & 0 & 2 & 9 \\ 0 & 1 & 5 & & 5 & 2 \\ & & 0 & 0 & & 7 \\ & & 0 & 0 & 0 & \end{array} \right] \end{array}$$

64   TRANSPORTATION NETWORKS

The maximum flow through the transportation network under observation is :

$$v^* = 5 + 13 = 9 + 2 + 7 = 18$$

Optimal flows are shown on Figure 38.

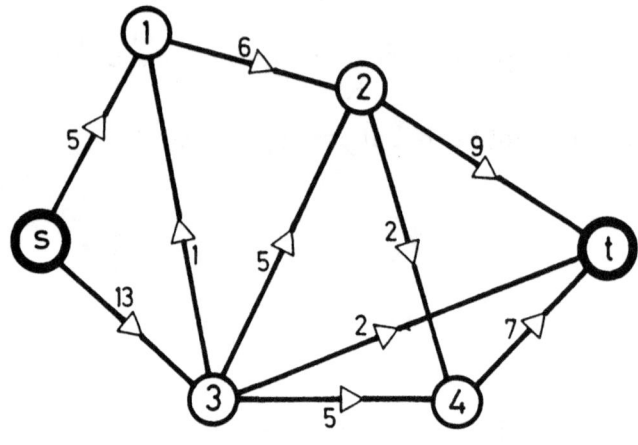

FIGURE 38. Maximum flows from nodes s to t.

3 Vehicle Routing Problems on Networks

Designing vehicle routes is a problem which is often encountered. Often, vehicles must call at a certain number of nodes in the transportation network or must go through specifically determined branches in the network. Collecting garbage, mail, cleaning streets, distributing newspapers, bread, milk, picking up schoolchildren, delivering packages, delivering mail by air, scheduling airplanes or buses on the network, scheduling plane crews and bus drivers for certain jobs, etc., are daily problems encountered by traffic and transportation experts.

Depending on whether vehicles must go along certain branches or call at certain nodes in the network, problems are differentiated into edge covering problems or node covering problems, respectively. These problems have been greatly studied in recent years and one example, the "traveling salesman" (the best-known node covering problem), has been the subject of hundreds of papers throughout the world.

In order to solve different variations of vehicle routing problems or crew scheduling problems, diverse techniques are applied, including dynamic programming and combinatorial programming (the branch-and-bound method). The heuristic procedure is also used to solve many problems of this type. In the majority of cases, the application of classical mathematical programming methods requires a great deal of computer work which rapidly increases with the increase in the number of nodes on the transportation. For this reason, many combinatorial problems are solved with heuristic procedures. These procedures, as a rule, do not require a lot of computer time, but on the other hand they do not guarantee that the optimal solution has been found. Still, heuristic procedures are finding a broad application in the field of engineering since they provide a solution in many cases which is relatively close to being optimal with a reduced amount of computer time. When problems are solved using classical methods of mathematical programming, certain limitations in the problem are often disregarded which significantly reduces the computer time and provides the optimal solution as well. However, the optimal solution found in this manner is no longer the optimal solution of the original problem. The application of certain heuristic procedures, dynamic programming, combinatorial programming or some other method primarily depends on the nature of the problem to be solved. There is no set rule as to when a certain method is to be used. Still, experience has shown that heuristic procedures provide satisfactory results when solving many problems dealing with vehicle routing, vehicle scheduling and driver (crew) scheduling on transportation networks.

VEHICLE ROUTING PROBLEMS ON NETWORKS   67

Problems concerning vehicle routing, determining the optimal position for the vehicle depot within the transportation system, crew planning, etc., belong to the class of so-called combinatorial problems. Combinatorial problems can be those dealing with sequences, assignments, choice making or any combination of these problems.[39]

For sequencing problems, there is usually a series of n elements whose objective functions reach an extreme value. For example, the traveling salesman has to visit n cities; determine the order in which he should visit them so that he covers the least distance (this is the so-called traveling salesman problem).

For problems dealing with assignments, there are usually n elements of a set which should be individually distributed onto n elements of another set so that the objective functions reach an extreme value. This can be used to distribute n drivers onto n buses as well.

For problems of choice making, m elements must be chosen from a set of n elements (m<n) so that the objective functions reach an extreme value.

As noted by Müller-Merbach,[39] most organizational problems on transportation networks are actually a combination of sequencing, assignment and choice making problems. The best example of this observation is in designing schedules on a transportation network which has air, bus and train traffic. If we wish to design bus, train or airplane schedules so that total passenger waiting time is minimized when changing types of transportation on the network, then we are presented with a problem containing a combination of sequencing, assignment and choice making.

The problem of planning airplane crews or bus drivers is also a combination of these three factors since a **crew** schedule usually tries to minimize costs and at the same time satisfy normal legal limitations regarding the length of the driver's work day, breaks or permitted daily number of flights by a pilot, etc.

The problem of scheduling m trucks or m airplanes (chartered) to n cities to be serviced (delivering or picking up merchandise) is also a combination of assignment and sequencing. It is clear that the schedule should attempt to minimize transportation costs while keeping the vehicle's capacity in mind (weight and/or volume), travel time (flying), service time in certain cities, etc.

Various modifications of these transportation problems appear in different branches of transportation, but all of them are very similar and are solved using the same or similar methodological process.

3.1. Vehicle routing on the network

One of the best-known and most-studied of the edge covering problems is that of the Chinese Postman, which consists of the following :

A postman is responsible for delivering mail to one part of a city. He starts his deliveries from a fixed node where the post office is located and must make the rounds of all the streets in his part of the city at least once during his working day, then return to the post office at the end of the day. The logical question to ask is : What route should the postman take so the total distance he travels is minimized and at the same time every street is visited at least once?

Translated into transportation network terminology, this question becomes : What is the shortest path to take through the network so that every branch is gone through at least once and we finish at the same node from which we started?

This problem was first studied by Leonhard Euler in 1736.

3.2. The Chinese Postman's problem on a nonoriented network

Let there be nonoriented network $G(N,A)$ with known lengths of all branches $l(i,j) > 0$, $(i,j) \in A$. As mentioned previously, $l(i,j)$ can represent costs, travel time, or anything else, depending on the problem. The Chinese Postman's problem can be formulated as follows :

Find a cycle which lets us go through every branch of network G at least once, for which :

$$\sum_{\text{all } (i,j) \in A} s_{ij} \cdot l(i,j) \longrightarrow \min$$

where s_{ij} is the number of passages through branch (i,j).

In order to solve this problem we must first define the terms Euler tour and Euler path.

A Euler tour is a cycle which goes through every branch of the network exactly once. A Euler path is a path which goes through every

70 TRANSPORTATION NETWORKS

branch of the network exactly once.

Euler stipulated that connected nonoriented network G contains a Euler tour (Euler path) if and only if network G contains exactly zero (exactly two) nodes with an odd degree. The proof of this theorem can be found in reference 29.

Figure 39 shows a network which does not contain odd degree nodes and therefore contains a Euler tour.

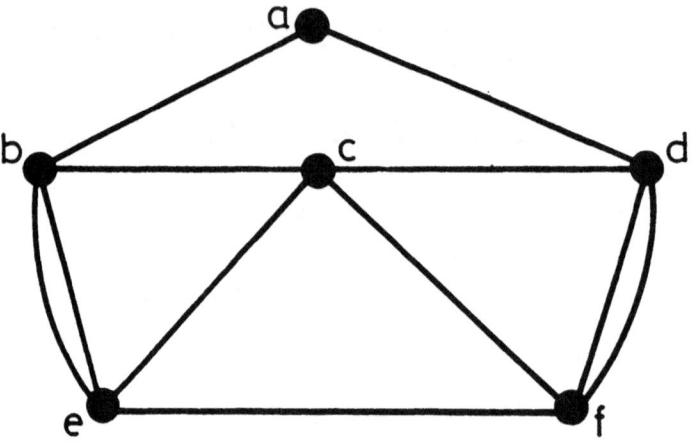

FIGURE 39. Network containing a Euler tour.

An example of a Euler tour through the network shown on Fig. 39 is shown by the cycle $(b, e, b, c, e, f, c, d, f, d, a, b)$.

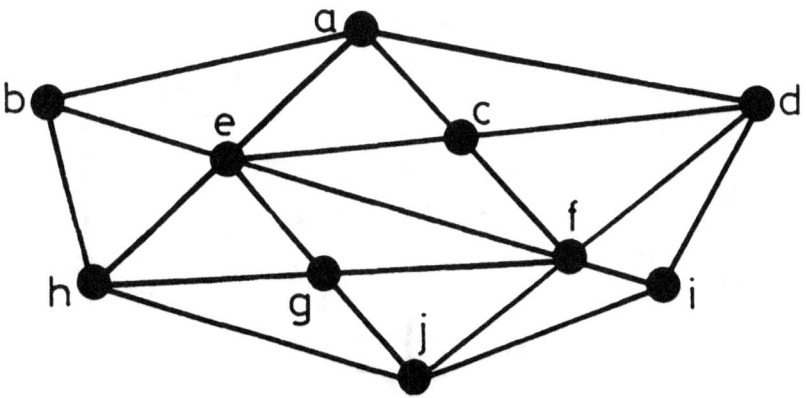

FIGURE 40. Network containing a Euler path.

In this network, only nodes b and i have an odd degree and for this reason the network contains a Euler path. One such path is as follows : (b, a, e, c, a, d, c, f, d, i, f, e, b, h, e, g, h, j, g, f, j, i).

We note once more that a Euler path, after going through every branch of the network, does not return to the starting node.

Figure 41 shows a network which has neither a Euler tour nor a Euler path, since nodes a, f, g and h have an odd degree.

It is clear that several different Euler tours can exist in connected nonoriented network $G(N,A)$ which does not have any odd degree nodes. However, all of these different Euler tours have an equal total length of :

$$\sum_{\text{all }(i,j)\in A} l(i,j)$$

We now return to the Chinese Postman problem

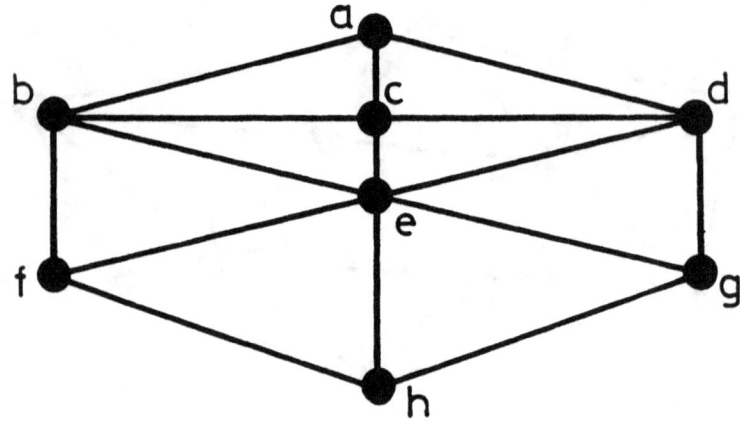

FIGURE 41. Network with neither a Euler path nor a Euler tour.

and solve it by finding the shortest tour going through every branch in connected nonoriented network G(N,A) at least once.

The procedure used to solve this problem consists of adding artificial branches parallel to the existing ones so that original network G is transformed into some new G'(N,A') in which we can make a Euler tour.

Artificial branches are only added to certain places in network G so that the odd degree nodes of network G are transformed into even degree nodes of network G'. The addition of artificial branches parallel to existing ones means that existing branches will be gone through twice in the final Chinese Postman's tour. The following claim is necessary to continue the procedure :

The number of odd degree nodes in a nonoriented network is always even. Let us study the network shown on Figure 42.

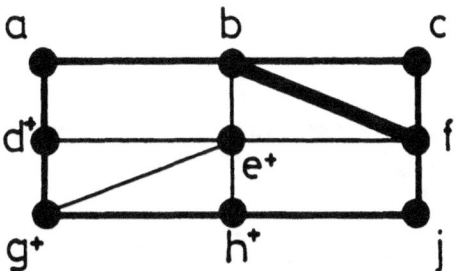

FIGURE 42. Nonoriented network with odd degree nodes.

A cross (+) is placed next to odd degree nodes. Let us look at branch (b,f). Since (b,f) connects two nodes, node b and node f, this branch will be counted when calculating the degree of node b, but it will also be counted when calculating the degree of node f. Since every branch in the network connects 2 nodes, we can conclude that the sum of all node degrees in the network is twice as large as the number of branches in the network. In other words, the sum of all node degrees in the network is always an even number. We denote this sum by S. The sum of even degrees in the network is denoted by P and the sum of odd degree nodes is denoted by N. It is clear that $N = S - P$. Since the difference between two even numbers is always even, the sum of odd degree nodes N is an even number. We denote by a_i the degree of node i which is an odd degree node. Since a_i is an odd number, we can write it as $a_i = 2m_i - 1$. We denote the total number of odd degree nodes in the network by k. It is clear that :

$$N = \sum_{i=1}^{k} a_i = \sum_{i=1}^{k} (2m_i - 1) = 2 \sum_{i=1}^{k} m_i - k$$

Finally, it is obvious that the number of odd degree nodes is :

$$k = 2 \sum_{i=1}^{k} m_i - N$$

an even number, which was to be shown.

The Chinese Postman's problem on a nonoriented network can be solved using the following algorithm :

Step 1 : Find all odd degree nodes in network $G(N,A)$. Let there be a total of k, with k being an even number.

Step 2 : Find $\frac{k}{2}$ pairs of these nodes so that the total branch length between these nodes iz minimized. In other words, find $\frac{k}{2}$ shortest paths between the k nodes.

Step 3 : For each $\frac{k}{2}$ pairs of nodes, add artificial branches parallel to existing branches on the shortest path between two nodes. New graph $G'(N,A')$ does not contain odd degree nodes.

Step 4 : Find a Euler tour in network $G'(N,A')$. This Euler tour is the optimal solution to the Chinese Postman problem in original network $G(N,A)$. The length of this optimal tour equals the sum of all branch lengths in network $G(N,A)$ and the length $\frac{k}{2}$ of the shortest path between studied $\frac{k}{2}$ pairs of nodes which were odd degree nodes in the original network.

Example : solve the Chinese Postman problem for a tour which starts and finishes in node a of the transportation network shown on Fig. 43.

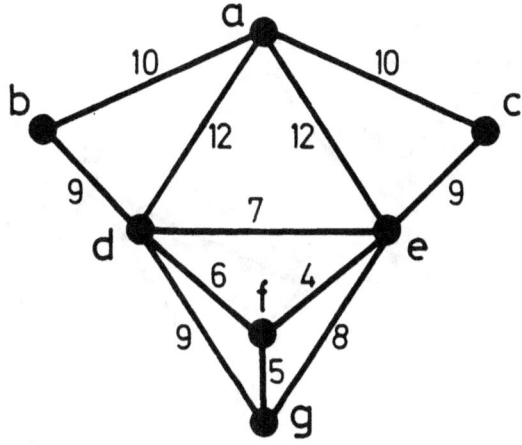

FIGURE 43. Nonoriented network for solving the Chinese Postman problem.

The figure shows 4 odd degree nodes : d, e, f and g. This means that the following are possible pairwise matchings of the odd degree nodes : d-e and f-g, d-g and f-e, and d-f and e-g. Figure 44 shows the three new networks which are obtained by introducing artificial branches between the nodes of all three possible pairwise matchings.

The shortest total length of the artificial branches is 12 which comes from matching d-e and f-g. This completes Steps 2 and 3 of our algorithm. We added one artificial branch parallel to branch (d,e) and one paralle to branch (f,g). Now we

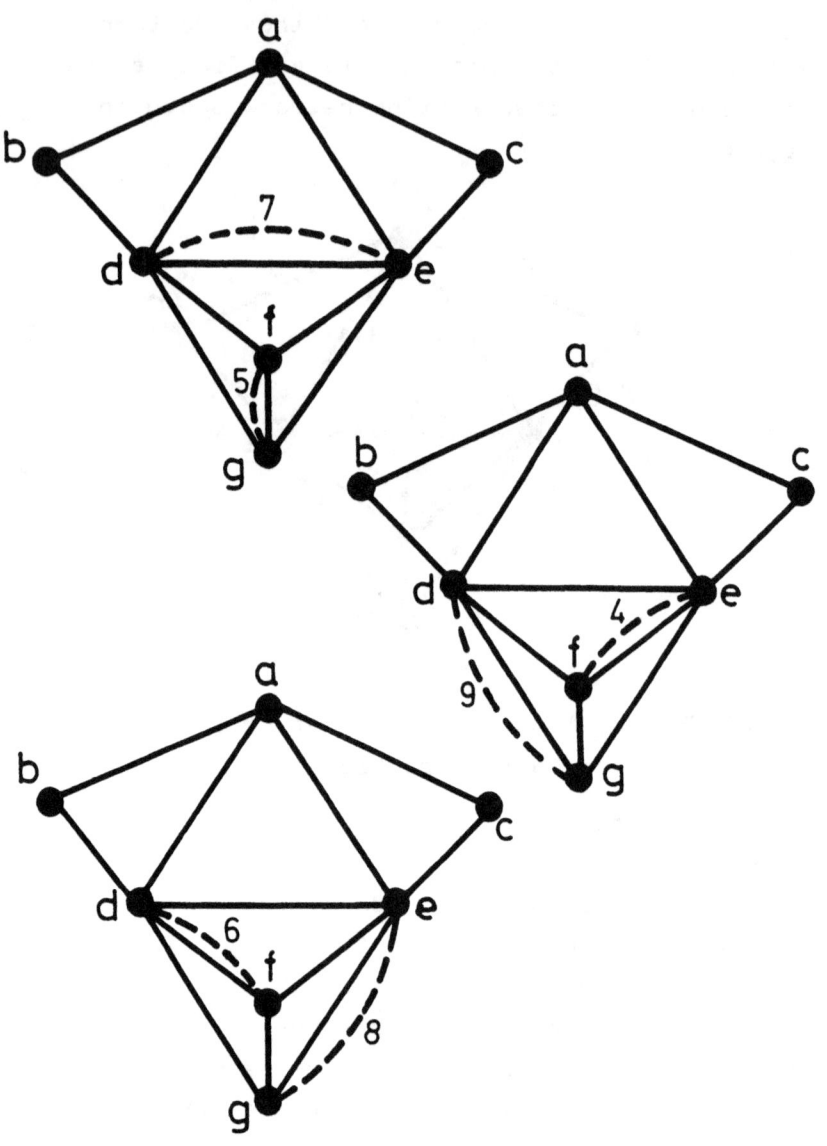

FIGURE 44. Three possible matchings of odd degree nodes.

VEHICLE ROUTING PROBLEMS ON NETWORKS   77

must find a Euler tour in the new network which starts at node a. The tour is as follows :

(a, d, g, e, f, g, f, d, e, d, b, a, c, e, a)

The total length of this tour is 113 with 101 belonging to the network branch lengths and 12 belonging to the artificial branches, which means that branches (d,e) and (f,g) are gone through twice.

Step 2 of the algorithm described is the most complex. The number of possible pairwise matchings of odd degree nodes rapidly increases with the number of nodes in the network, although for the majority of practical problems this matching can be done without using a computer.

Experience shows that very good pairwise marching can be obtained by simply studying a geographical map. As mentioned by Larson and Odoni[29], pairwise matching without the use of a computer is greatly facilitated by the fact that in pairwise matchings with a minimal total path length between odd degree nodes, there cannot be two shortest paths which contain a common branch.

In order to further explain this statement, let us take a look at Figure 45.

We assume that nodes r, s, t and q are odd degree nodes and that the shortest paths between nodes r-t and nodes s-q have a common branch (i,j). When pairwise matching nodes with the minimum branch length, node pairs r-t and s-q cannot be included since this matching can be replaced by node pairs r-s and t-q whose total branch length of 2 l(l,j) is less than the total branch length

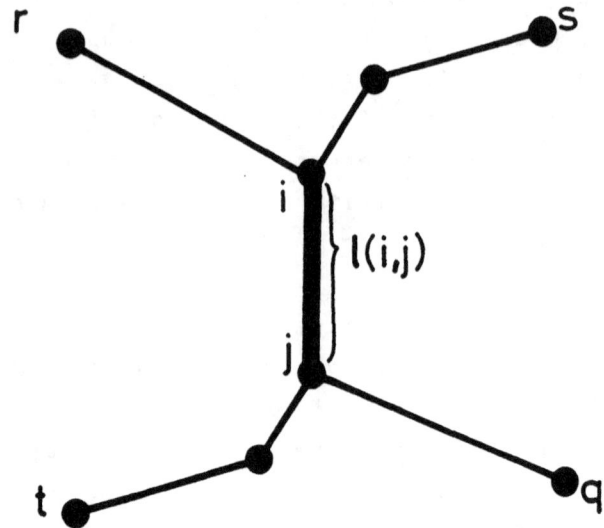

FIGURE 45. One method of solving pairwise matching for odd degree nodes.

of node pairs r-t and s-q. This demonstrates that when pairwise matching minimum total branch lengths there cannot be two shortest paths which contain a common branch. Bearing this fact in mind, when pairwise matching without a computer, a large number of theoretically possible odd degree node matchings should be eliminated so that odd degree nodes should be paired with other odd degree nodes in their immediate vicinity.

3.3. The Chinese Postman problem on an oriented network

The Chinese Postman problem on an oriented network was solved by E. Beltrami and L. Bodin[4] in 1974. To do this they used the following version of Euler's theorem which refers to oriented networks:

A connected oriented network contains a Euler tour if and only if the in-degree of every node equals the out-degree of the same node.

The proof of this theorem is completely analogous to that of Euler's theorem for a non-oriented network. In order to solve the Chinese Postman problem for an oriented network, Beltrami and Bodin first defined node "polarity" as the difference between the in-degree and out-degree of the node. Nodes n_j whose in-degree is greater than their out-degree are called supply nodes. We denote the polarity of these nodes by s_j. Nodes whose out-degree is greater than their in-degree are denoted by m_k and are called demand nodes. Their polarity is designated by d_k.

Step 1 : Find all supply nodes and all demand nodes.
Step 2 : Find the shortest paths d_{jk} from all nodes n_j to all nodes m_k.
Step 3 : Solve the linear programming problem in order to find the optimal pairwise matching of supply nodes with demand nodes. This problem reads as follows :

$$z = \sum_j \sum_k d_{jk} x_{jk} \longrightarrow \min$$

$$\sum_k x_{jk} = s_j \quad \text{for all } j$$

$$\sum_j x_{ij} = d_k \quad \text{for all } k$$

$$x_{jk} \geq 0$$

80 TRANSPORTATION NETWORKS

Step 4 : For every $x_{jk} > 0$ obtained as a solution to the linear programming problem we add x_{jk} artificial paths parallel to the shortest path from n_j to m_k. New network G' obtained in this manner takes on the polarity of all its nodes and equals zero.

Step 5 : Find a Euler tour for network G'. This tour solves the Chinese Postman problem on an oriented network.

E x a m p l e : apply the algorithm described above to the network shown in Fig. 46 and find a tour which starts and finishes at node b.

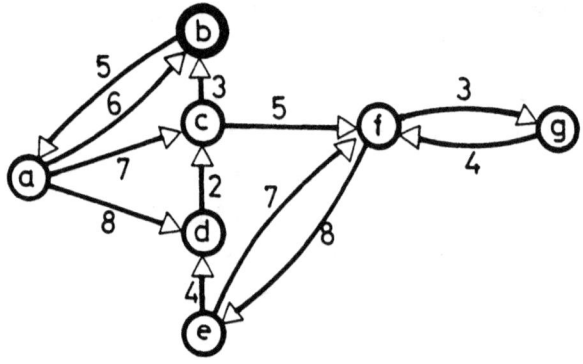

FIGURE 46. Oriented network for solving the Chinese Postman problem.

We first calcualte the in-degree, out-degree and polarity of all nodes in the network and thereby determine whether they belong to the set of supply nodes S or the set of demand nodes D. The results are given in Table III.

TABLE III. In-degrees, out-degrees, polarity and set nodes belong to on Fig. 46.

Node i	In-degree	Out-degree	Polarity	Set it belongs to
a	1	3	-2	D
b	2	1	1	S
c	2	2	0	
d	2	1	1	S
e	1	2	-1	D
f	3	2	1	S
g	1	1	0	

Sets S and D contain the following nodes :

$$S = \begin{bmatrix} b, d, f \end{bmatrix} \qquad D = \begin{bmatrix} a, e \end{bmatrix}$$

The shortest paths (see Fig. 46) from the nodes of set S to the nodes of set D are :

$$d_{b,a} = 5$$
$$d_{b,e} = 5 + 7 + 5 + 8 = 25$$
$$d_{d,a} = 2 + 3 + 5 = 10$$
$$d_{d,e} = 2 + 5 + 8 = 15$$
$$d_{f,a} = 8 + 4 + 2 + 3 + 5 = 22$$
$$d_{f,e} = 8$$

Our transportation problem now reads :

$$5x_{b,a} + 25x_{b,e} + 10x_{d,a} + 15x_{d,e} + 22x_{f,a} + 8x_{f,e} \longrightarrow \min$$

$$x_{b,a} + x_{b,e} = 1$$

82 TRANSPORTATION NETWORKS

$$x_{d,a} + x_{d,e} = 1$$
$$x_{f,a} + x_{f,e} = 1$$
$$x_{b,a} + x_{d,a} + x_{f,a} = 2$$
$$x_{b,e} + x_{d,e} + x_{f,e} = 1$$

We denote $x_{b,a}$ by x_1, $x_{b,e}$ by x_2, etc. Now we get :

$$F = 5x_1 + 25x_2 + 10x_3 + 15x_4 + 22x_5 + 8x_6 \rightarrow \min$$

$$x_1 + x_2 = 1 \quad \ldots\ldots\ldots \text{(I)}$$
$$x_3 + x_4 = 1 \quad \ldots\ldots\ldots \text{(II)}$$
$$x_5 + x_6 = 1 \quad \ldots\ldots\ldots \text{(III)}$$
$$x_1 + x_3 + x_5 = 2 \quad \ldots\ldots\ldots \text{(IV)}$$
$$x_2 + x_4 + x_6 = 1 \quad \ldots\ldots\ldots \text{(V)}$$

With the help of constraints (I,), (II) and (III) we denote x_1, x_3 and x_5 through x_2, x_4 and x_6. We now have :

$$x_1 = 1 - x_2$$
$$x_3 = 1 - x_4$$
$$x_5 = 1 - x_6$$

By adding we get :
$$(x_1 + x_3 + x_5) = 3 - (x_2 + x_4 + x_6)$$
Since, (based on constraint V)
$$x_2 + x_4 + x_6 = 1$$
This is :
$$x_1 + x_3 + x_5 = 3 - 1 = 2$$

We have obtained a constraint (IV) which is not independent as it is actually a combination of constraints (I), (II), (III) and (V). Our linear programming problem now reads:

$$F = 5x_1 + 25x_2 + 10x_3 + 15x_4 + 22x_5 + 8x_6 \longrightarrow \min$$

$$x_1 + x_2 = 1$$
$$x_3 + x_4 = 1$$
$$x_5 + x_6 = 1$$
$$x_2 + x_4 + x_6 = 1$$

Since n (number of variables) - m (number of constraints) = 6 - 4 = 2, we can solve this problem graphically. We take x_1 and x_3 as non-basic variables and express the other (basic) variables through x_1 and x_3. This gives us:

$$x_2 = 1 - x_1$$
$$x_4 = 1 - x_3$$
$$x_5 = 2 - x_1 - x_3$$
$$x_6 = x_1 + x_3 - 1$$
$$F = -34x_1 - 19x_3 + 76 \longrightarrow \min$$

The function's minimum is for the same values as the function $F' = -34x_1 - 19x_3$. Using the well-known graphic procedure, we obtain Figure 47 on the next page.

$$x_1 \geq 0; \quad x_3 \geq 0$$
$$x_2 \geq 0; \quad 1 - x_1 \geq 0; \quad x_1 \leq 1$$
$$x_4 \geq 0; \quad 1 - x_3 \geq 0; \quad x_3 \leq 1$$

$$x_5 \geq 0; \quad 2 - x_1 - x_3 \geq 0; \quad x_1 + x_3 \leq 2$$
$$x_6 \geq 0; \quad x_1 + x_3 - 1 \geq 0; \quad x_1 + x_3 \geq 1$$

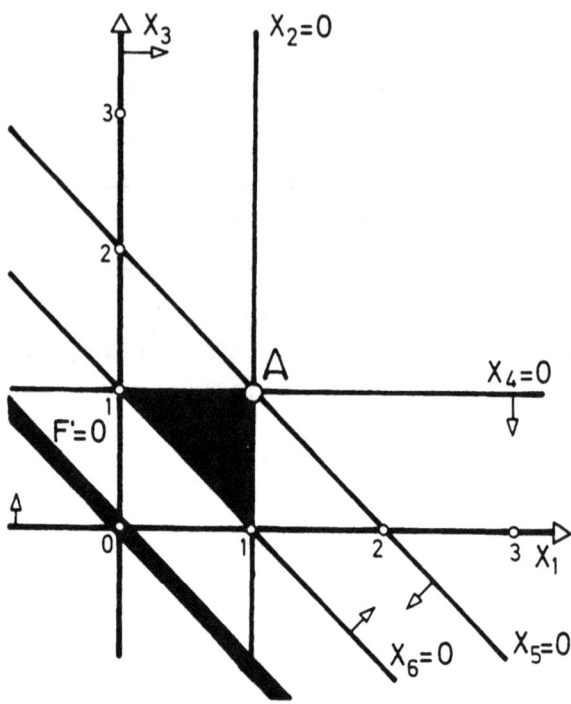

FIGURE 47. Graphic procedure for solving the Chinese Postman problem in an oriented network.

The minimum of function F' is in point A, making the optimal solution :

$$x_1 = 1$$
$$x_2 = 0$$
$$x_3 = 1$$
$$x_4 = 0$$
$$x_5 = 0$$
$$x_6 = 1$$

The value of function F for this solution is 23. So the following is therefore obtained :

$$x_1 = x_{b,a} = 1$$
$$x_3 = x_{d,a} = 1$$
$$x_6 = x_{f,e} = 1$$

This means that a parallel artificial path must be added to the paths between b and a, d and a and f and e. The new network in which the polarity of all nodes equals zero looks like this(Fig.48):

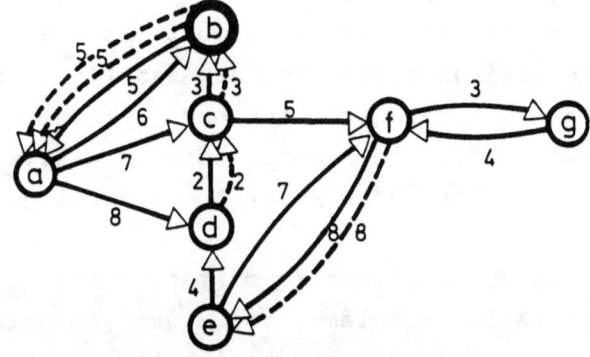

FIGURE 48. Chinese Postman network where node polarity equals zero.

The dotted lines denote artificial branches between nodes. The total length of the added branches is 23. The new network contains a Euler tour which starts and finishes in node b :
(b, a, b, a, c, b, a, d, c, f, e, f, g, f, e, d, c, b).

3.4. The traveling salesman problem

When distributing, picking up or carrying out some other service in which certain nodes must be called on, vehicle routing problmes become node covering problems. The most important and best-known of all node covering problmes is that of the traveling salesman which can be defined as follows:

Find the shortest itinerary which starts in a specific node, goes through all other nodes at least once and finishes in the starting node.

The traveling salesman problem can also contain the stricter requirement that each node be gone through exactly once. Obviously, this requirement is made under the assumption that such a tour can be found.

When the traveling salesman must visit each node in the transportation network exactly once and then return to his starting node, traveling the shortest possible distance, this is usually called the classical problem of the traveling salesman.

We should mention that in different traffic and transportation problems, the traveling salesman can represent airplanes, boats, trucks, buses, crews, etc. Vehicles visiting nodes can deliver or pick up merchandise or passengers, or, for example, simultaneously pick up and deliver merchandise. One of the most important papers devoted to the problem of the traveling salesman was written by Pierce in 1969.[46] Similar to Pierce, we denote by D_o the depot from which vehicles start and finish, and by $D_1, D_2, \ldots, D_i, \ldots, D_m$ the nodes to be visited. We denote the transport time between node i and node j by $c(i,j)$. Depending on the specific problem to be solved, time can be replaced by the distance between nodes i and j, transport costs or some other quantity. We would note that $c(j,i)$ can, but does not have to equal the quantity $c(i,j)$. We denote by P_i the time needed to carry out the service in node D_i and $D_\emptyset$ denotes the node to which the vehicle returns after providing service to all m nodes. Node $D_\emptyset$ is actually starting node D_o from which the vehicle started.

Our problem is composed of finding the optimal route $D_o, D_{i_1}, D_{i_2}, \ldots, D_{i_m}, D_\emptyset$ for which :

$$A\theta = P_o + \sum_{k=1}^{\emptyset} \left[c(i_{k-1}, i_k) + P_{i_k} \right] \rightarrow \min$$

where $\theta = (o, i_1, i_2, \ldots, i_m, \emptyset)$ is the permutation of nodes to be visited for which the total time spent traveling is minimized. The time needed to prepare the vehicle (fill it up, load it) in depot D_o is denoted by P_o. Since the total time needed to provide service to the nodes is a constant number, we also have :

88 TRANSPORTATION NETWORKS

$$p_o + \sum_{k=1}^{\emptyset} p_{i_k} = \text{const}$$

and our problem can also be read as :

Find the permutation $\bullet = (o, i_1, i_2, \ldots, i_m, \emptyset)$ of elements 1, 2,..., m for which :

$$A_o = \sum_{k=1}^{\emptyset} c\,(i_{k-1}, i_k) \longrightarrow \min$$

It has already been mentioned that this type of problem can be solved either by exact mathematical methods or by using heuristic procedures. It has been shown that the best suited mathematical method for solving this problem is combinatorial programming. As mentioned by J.F. Pierce,[46] the combinatorial programming method is also known as the branch-and-bound method and reliable heuristic programming. In recent years the term "branch-and-bound method" has primarily been used, and we will henceforth use this term.

Let a vehicle leave node 0 and let it have to visit nodes 1, 2, 3 before returning to node 4. Node 4 is actually starting node 0. These nodes can be visited in 6 different ways. Figure 49 shows the 6 different ways to visit nodes 1, 2 and 3 with a return to starting node 0 (node 4).

The possible solutions to visiting nodes 1, 2 and 3 after starting off at node 0 are presented in the shape of a tree. Each path in the tree leading from node 0 to node 4 is one of the possible solutions to the problem. The tree was formed by starting from node 0 and branching into three

VEHICLE ROUTING PROBLEMS ON NETWORKS 89

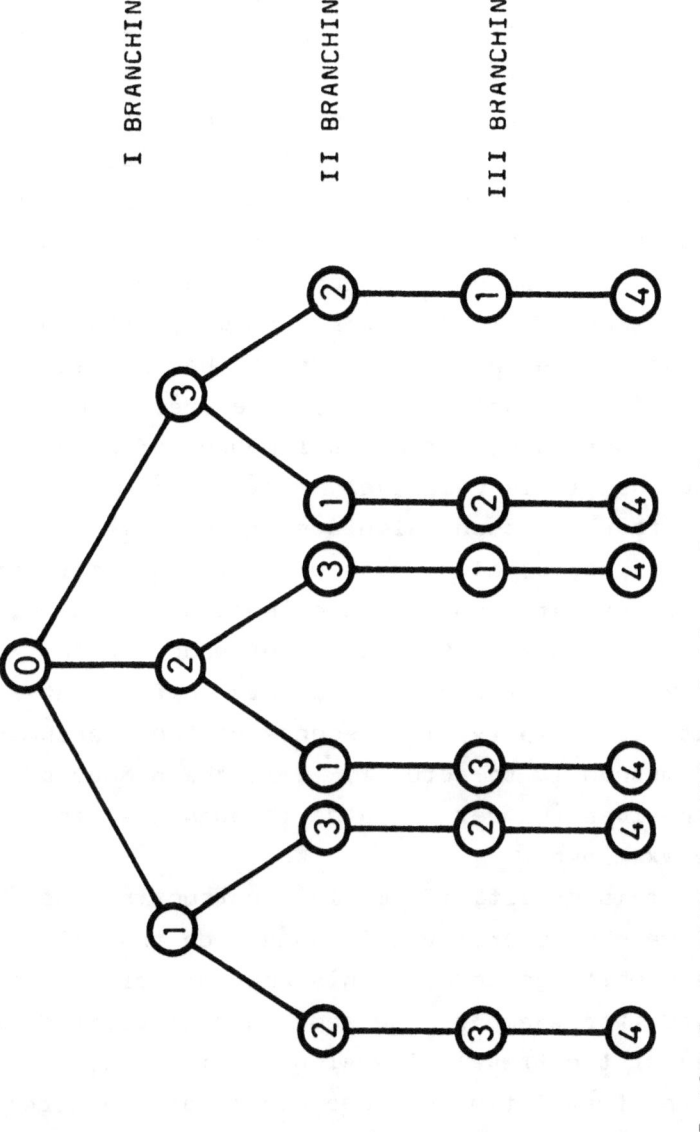

FIGURE 49. Tree with possible solutions for visiting nodes.

branches towards nodes 1, 2 and 3 (the three possible first visits). We continue branching from these nodes and examine all possible ways of continuing our trip. This process of branching when generating all possible solutions resulted in the word "branch" in the process name. The tree shown on Fig. 49 was very easy to construct since the three nodes to be visited generated a relatively small number of alternatives. However, with an increase in the number of nodes to be visited, the number of alternative paths in the tree also rapidly increases. For a small number of nodes in the network it is relatively easy to branch the entire tree and then calculate travel time (or length or costs) along each path. It is clear that the optimum path contains the least travel time. With an increase in the number of nodes in the network, this process becomes very inefficient and unsuitable to apply. The essence of the branch-and-bound method is to actually limit the number of alternatives (number of paths through the tree) to be examined.

To this effect, if, while constructing the tree, we should discover in some node that all paths R which go through this node are either prohibited (for example, the same node is visited twice) or the travel time along it is greater than the travel time on some other path we already know of, then there is no need to continue branching from that node since the solution we receive will be either prohibited or worse than the solution we already have. Let us assume that A_j

and A_{θ_k} are total travel times along paths θ_j and θ_k. Path θ_j is better than path θ_k if :

$$A_{\theta_j} \leq A_{\theta_k}$$

We denote by θ^* the path with the least total travel time discovered in the transportation network to now. Limiting the number of alternatives to be examined is relevant to the fact that we are only interested in finding paths θ for which :

$$A_\theta < A_{\theta^*}$$

Quantity A_{θ^*} is called the upper bound of the optimal solution.

Let us assume that we are in some node on the k-th level of the tree in the process of searching for the optimal solution, and we designate by T_k the total travel time from the starting node to the node where we are presently located. If :

$$T_k \geq A_{\theta^*}$$

there is no need for further branching from this node since the travel time from the starting node to the node on the k-th level of the tree is already greater than (or equal to) the total travel time discovered to the present on a path we already know of (the path with the shortest travel time discovered to the present). This enables considerable savings in the number of branchings to be made.

We now denote by B_k the least possible travel time from the node on the k-th level to the end. Quantity B_k is often called the lower bound. If we are sure that the travel time from the node on the k-th level to the end is greater than or equal

to quantity B_k, then we can further decrease the number of alternatives to be examined by the following test. If :
$$T_k + B_k \geq A_{\ominus *}$$
then there is no need for further branching since total travel time would certainly be greater than or equal to quantity $T_k + B_k$, i.e. greater than or equal to $A_{\ominus *}$. Therefore, we have established that there are two tests which can essentially reduce the number of alternatives to be examined. The basic problem when carrying out the second test is that of calculating the value of B_k. The method which accomplishes this was proposed by Little in 1963 and is commonly called the "matrix reduction" method. Details of this method can be found in reference 46.

E x a m p l e : a vehicle starts from node 0, services nodes 1, 2, 3 and 4 and returns to node 0 (node 5). Servicing time (loading, unloading, filling) in nodes 0, 1, 2, 3 and 4 is given in Table IV.

TABLE IV. Servicing time.

Node D_i	0	1	2	3	4
Servicing time p_i	4	2	3	1	2

Travel time between nodes is given in the matrix:

$$\begin{array}{c} \\ 0 \\ 1 \\ 2 \\ 3 \\ 4 \end{array} \begin{array}{ccccc} 1 & 2 & 3 & 4 & 5 \end{array} \\ \left[\begin{array}{ccccc} 4 & 12 & 12 & 16 & \infty \\ \infty & 3 & 6 & 2 & 1 \\ 4 & \infty & 2 & 3 & 8 \\ 9 & 6 & \infty & 7 & 4 \\ 1 & 7 & 4 & \infty & 6 \end{array} \right]$$

VEHICLE ROUTING PROBLEMS ON NETWORKS   93

Using the method described, determine the vehicle's itinerary so that it services nodes 1, 2, 3 and 4 in the least possible time.

We can go from node 0 to any of the four nodes to be serviced. Therefore, we branch from node 0 to all four of the nodes (Fig. 50).

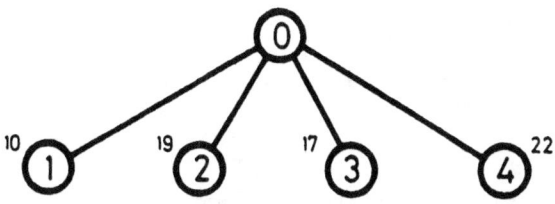

FIGURE 50. Branching from node 0 to nodes 1, 2, 3 and 4 (first branching).

The time needed to service the starting node is $p_0 = 4$. Travel time from node 0 to node 1 is $c(0,1) = 4$, and the time needed to service node 1 is $n_1 = 2$. The time needed from the beginning of service in node 0 to the end of service in node 1 is:

$$p_0 + c(0,1) + p_1 = 4 + 4 + 2 = 10$$

This time is noted next to node 1 on Fig. 50. It is clear that the travel time along all paths which first go through node 1 must be greater than or equal to 10.

Fig. 50 also shows the time needed from the beginning of service in node 0 to the end of service in nodes 2, 3 and 4 next to each node. We now continue branching from node 1 since path (0,1) offers the least total travel time for now. So we continue branching from node 1 towards nodes

94   TRANSPORTATION NETWORKS

2, 3 and 4 (Fig. 51).

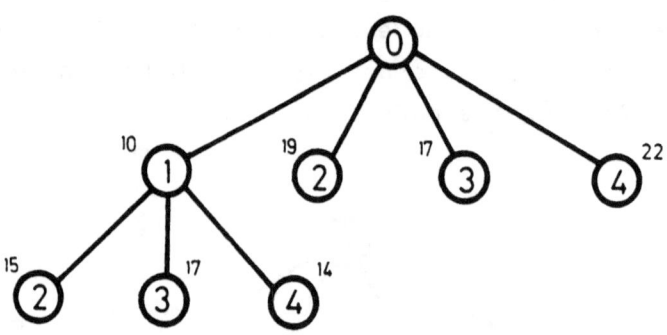

FIGURE 51. Second branching.

Travel time from node 1 to node 2 is c(1,2) =3, and the time needed to service node 2 is p_2 = 3. The total time needed from the beginning of service in starting node 0 to the end of service in node 2 is :

10 + 3 + 2 = 15

This time is noted next to node 2 of Fig. 51. Nodes 3 and 4 on Fig. 51 also contain the time needed from the beginning of service in the starting node to the end of service in these nodes. The time associated with each node in our tree therefore represents the cumulative travel time (with completion of service) along the path leading from the starting node to the node in question.

We continue branching from node 4 on the path (0, 1, 4) (Fig. 52).

The next two branchings are shown on Figs. 53 and 54, respectively.

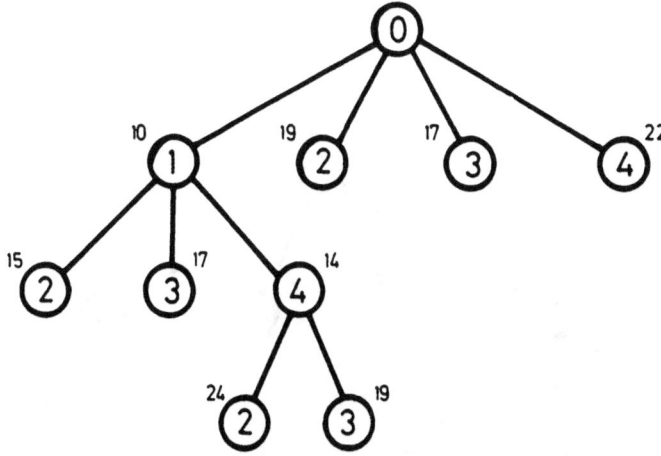

FIGURE 52. Third branching.

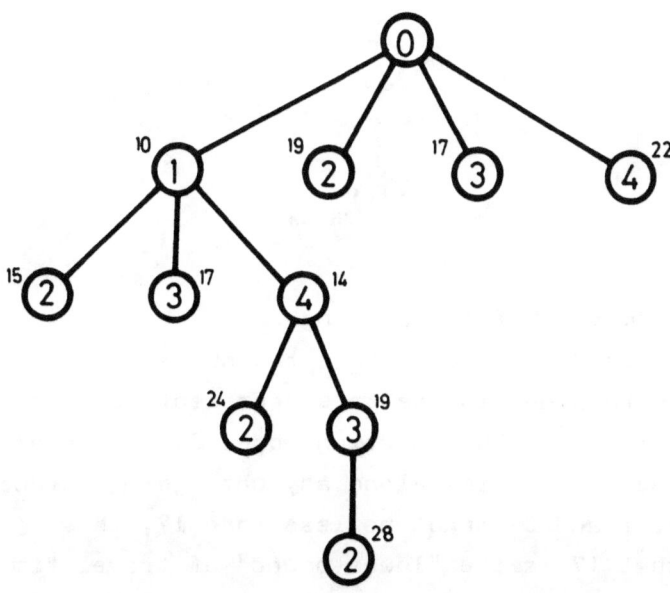

FIGURE 53. Fourth branching.

96 TRANSPORTATION NETWORKS

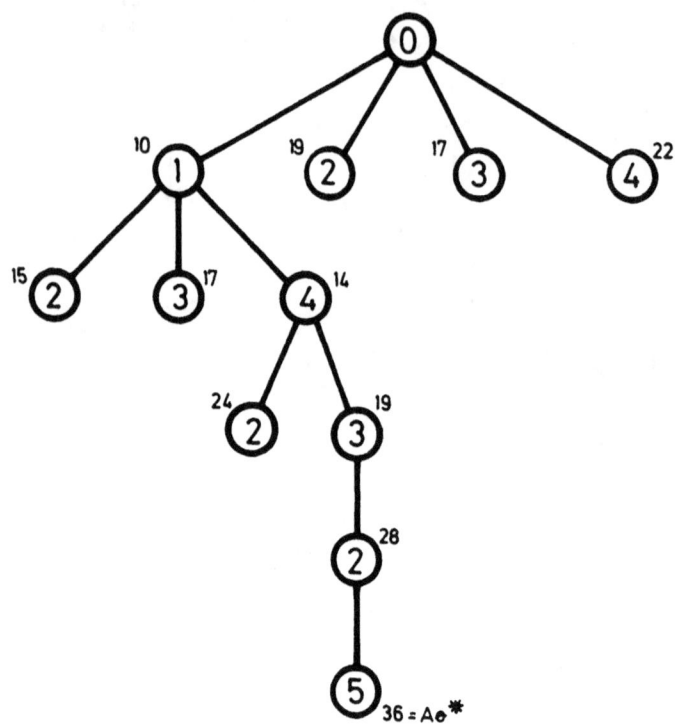

FIGURE 54. Fifth branching.

We already mentioned that the numbers associated with nodes in the tree represent cumulative travel time from the starting node. So, for example, since the travel time along any path going through nodes 0, 1 and 3 cannot be less than 17, this means that 17 is the "lower bound" of travel time for all paths which go through nodes 0, 1 and 3.

As can be seen from Fig. 54, we have construct-

ed one path in the tree, path $(0, 1, 4, 3, 2, 5)$.
The total travel time (including service) along
this path is 36. It is clear that we must continue
searching for a path whose travel time is less
than or equal to 36. But for the time being, the
least total travel time is 36, since we have only
researched path $(0, 1, 4, 3, 2, 5)$. For this
reason we put 36 as the upper bound of the optimal
solution $A_{\theta*}$. Regardless of the travel time along
paths yet to be examined, we are sure that our
optimal solution cannot be greater than 36.

We continue branching from the last node in
which we stopped branching when constructing path
$(0, 1, 4, 3, 2, 5)$. So we return from node 5 to
node 2, then node 3, then node 4 and from there go
down to node 2 on the path $(0, 1, 4, 2)$. This
backtracking is shown on Fig. 55 by dotted lines.

Total travel time associated with node 2 on
the path $(0, 1, 4, 2)$ is $T_3 = 24$ (index 3 indicates
that this node is on the third level of branching).
Since :

$$24 = T_3 < A_{\theta*} = 36$$

the first test to limiting further branching has
been successfully passed. In order to continue
branching from this node, we first establish B_3,
i.e. the least possible travel time (including
service) from this node to the end. Two more trips
must be made to reach the end. From the starting
travel time matrix we see that the shortest possible
travel time from from node 2 to any other node is
2. The shortest possible service time in any node
is 1. So we have :

98   TRANSPORTATION NETWORKS

$$B_3 = 2 \cdot 2 + 1 = 5$$

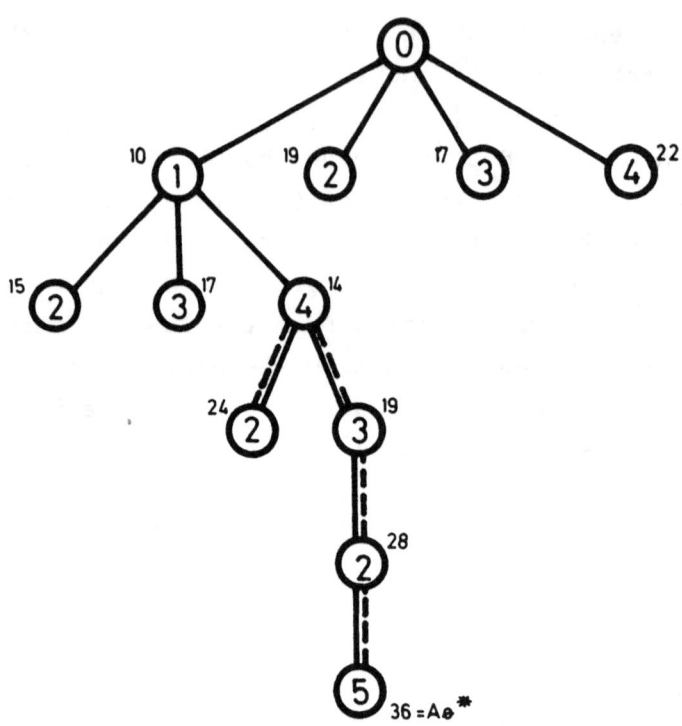

FIGURE 55. First backtracking.

And since :

$$T_3 + B_3 = 24 + 5 = 29 < 36 = A_{\theta *}$$

the second test has also been successfully passed which means that we can continue branching from node 2 on the path $(0, 1, 4, 2)$. From this node we can only continue to node 3 (Fig. 56).

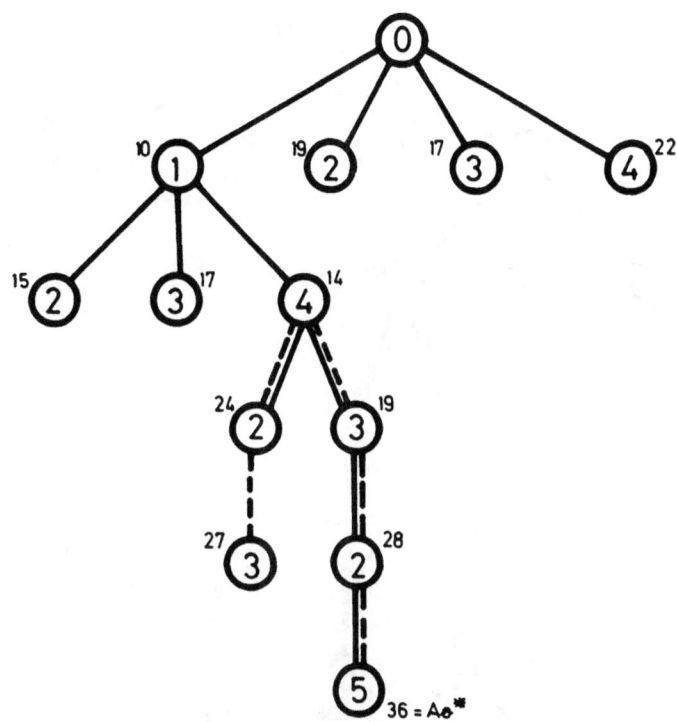

FIGURE 56. Sixth branching.

Since :
$$T_4 = 27 < 36 = A_{\theta*}$$
$$T_4 + B_4 = 31 < 36 = A_{\theta*}$$
we can continue branching from this node. Total travel time (with service) along this path equals 31 (Fig. 57).

100 TRANSPORTATION NETWORKS

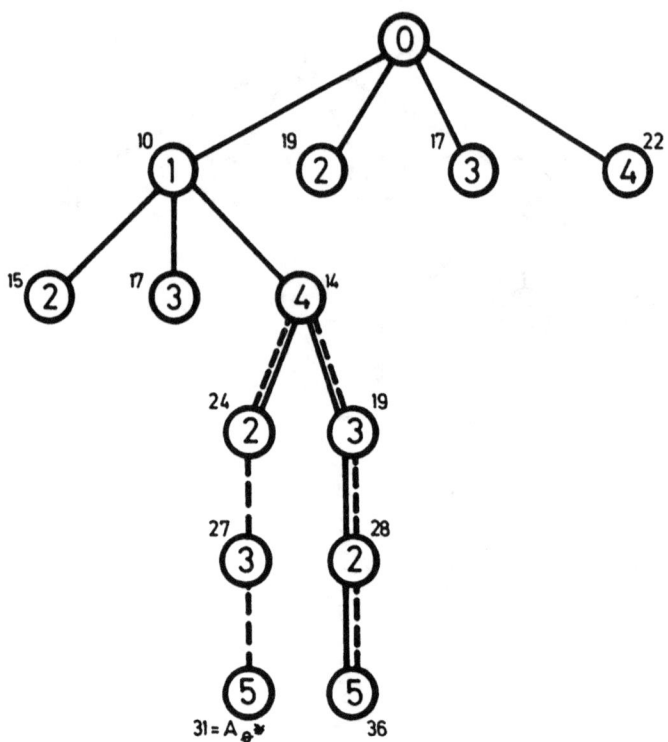

FIGURE 57. Seventh branching.

Since total travel time along path ө = $(0, 1, 4, 2, 3, 5)$ is 31 and since it is less than the upper bound of the optimal solution, i.e. since

$$31 = A_\theta < A_{\theta*} = 36$$

the new upper bound of the optimal solution now becomes 31. So we now have :

$$A_{\theta*} = 31$$

VEHICLE ROUTING PROBLEMS ON NETWORKS 101

To continue, we now return up path $(0, 1, 4, 2, 3, 5)$ from node 5 to node 3, node 2, node 4 and from node 1 we go down to node 3 on the path $(0, 1, 3)$. We continue branching from this node (Fig. 58).

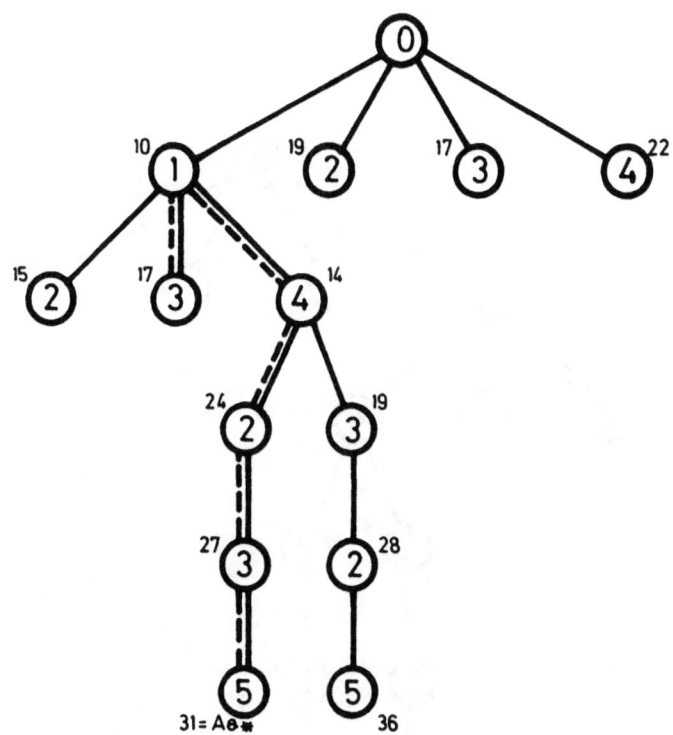

FIGURE 58. Second backtracking.

Node 3 on the path $(0, 1, 3)$ is on the second level. For this node we have :

102 TRANSPORTATION NETWORKS

$$T_2 = 17 < 31 = A_{\Theta*}$$

$$B_2 = 3 \cdot 4 + 2 \cdot 1 = 14$$

$$T_2 + B_2 = 31 = 31 = A_{\Theta*}$$

This means that if we continue branching from this node, we will receive a total travel time of

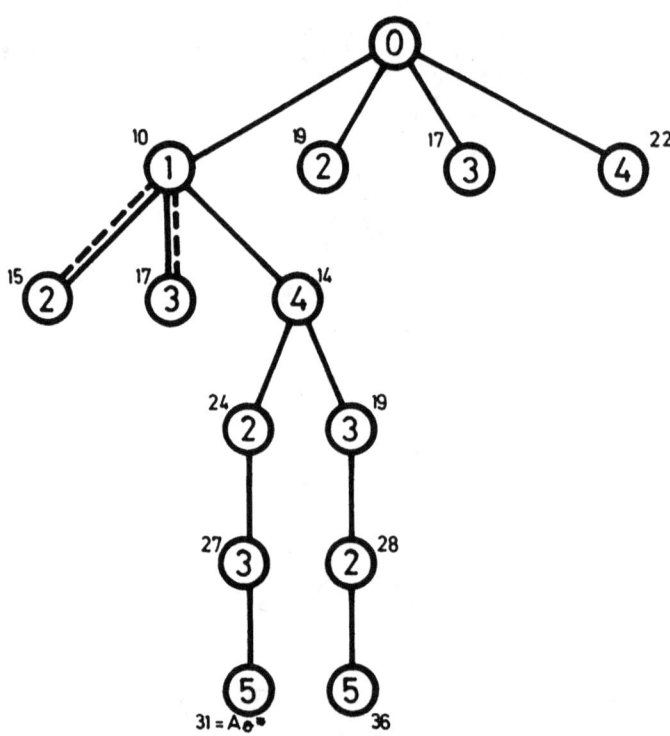

FIGURE 59. Third backtracking,

VEHICLE ROUTING PROBLEMS ON NETWORKS 103

least equal to the travel time we already have. So there is no need to continue branching from this node. We now return to node 1 and drop down to node 2 (Fig. 59). Node 2 on path $\left(0, 1, 2\right)$ is the next node from which branching continues.

We can continue branching from node 2 since :

$$T_2 = 15 < 31 = A_{\theta *}$$
$$B_2 = 3 \cdot 2 + 2 \cdot 1 = 8$$
$$T_2 + B_2 = 23 < 31 = A_{\theta *}$$

We now come to nodes 3 and 4 (Fig. 60).

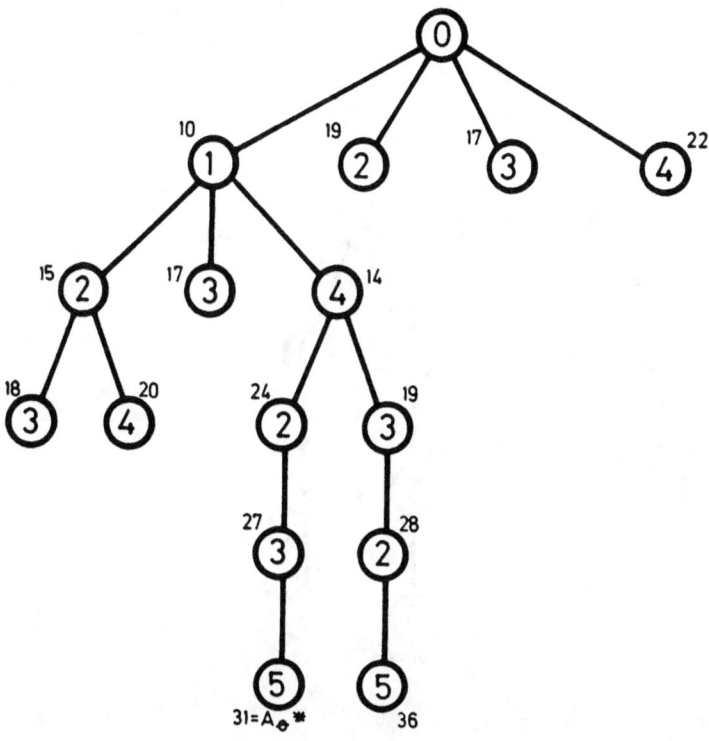

FIGURE 60. Eighth branching.

104 TRANSPORTATION NETWORKS

For node 3 we have :

$$T_3 = 18 < 31 = A_{\Theta*}$$
$$B_3 = 2 \cdot 4 + 1 \cdot 1 = 9$$
$$T_3 + B_3 = 27 < 31 = A_{\Theta*}$$

which means that we can continue branching from this node. We arrive at node 4 (Fig. 61).

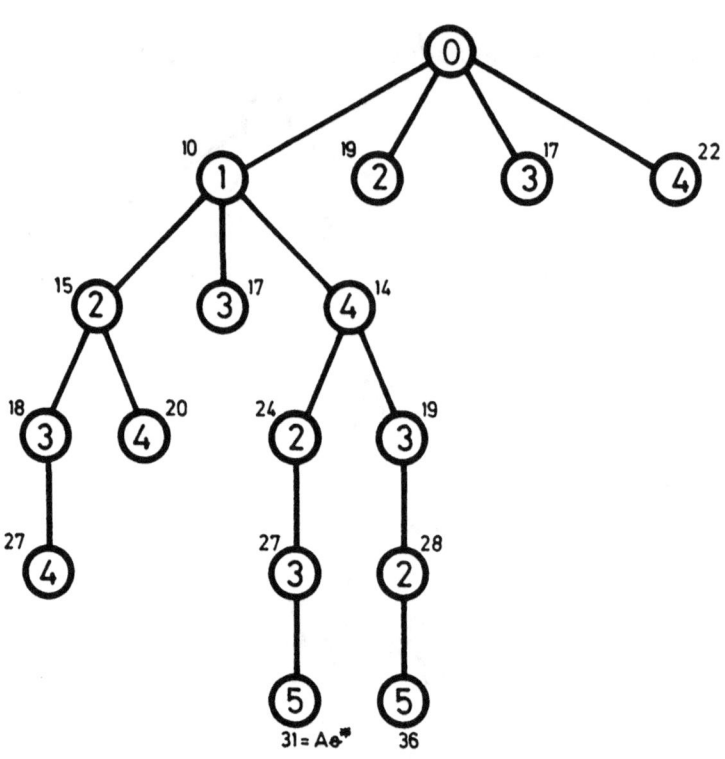

FIGURE 61. Ninth branching.

We can continue branching from node 4 since :
$$T_4 = 27 < 31 = A_{\theta*}$$
$$T_4 + B_4 = 29 < 31 = A_{\theta*}$$
We have now constructed path $\theta = (0, 1, 2, 3, 4, 5)$ (Fig. 62).

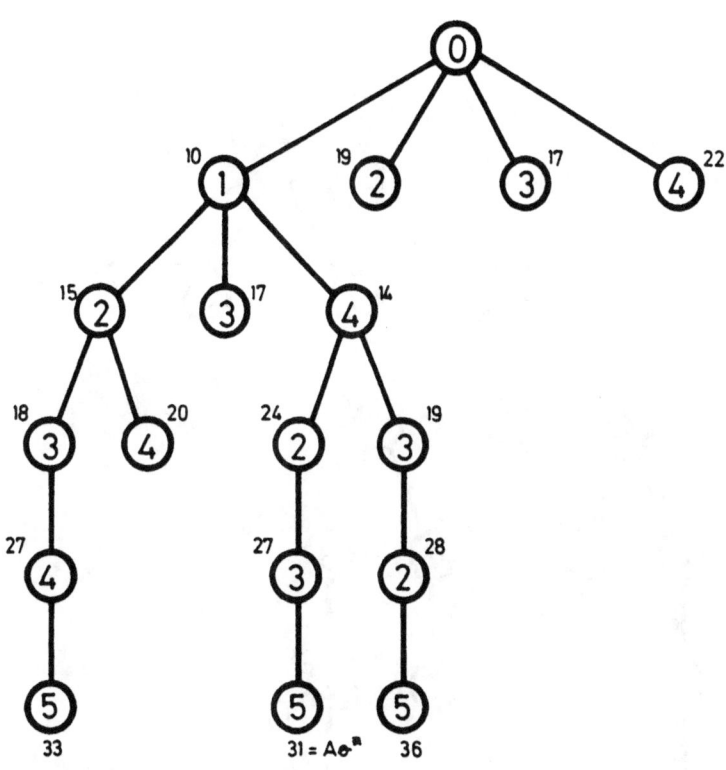

FIGURE 62. Tenth branching.

Since travel time A_θ along this path is greater than the upper bound of the optimal solution, i.e.

106 TRANSPORTATION NETWORKS

since :
$$A_\Theta = 33 > 31 = A_{\Theta*}$$
our upper bound for the optimal solution remains unchanged.

In order to continue branching we return along path $(0, 1, 2, 3, 4, 5)$ from node 5 to node 4, to node 3 and to node 2 from which we drop down to the third level of branching in node 4 (Fig. 63).

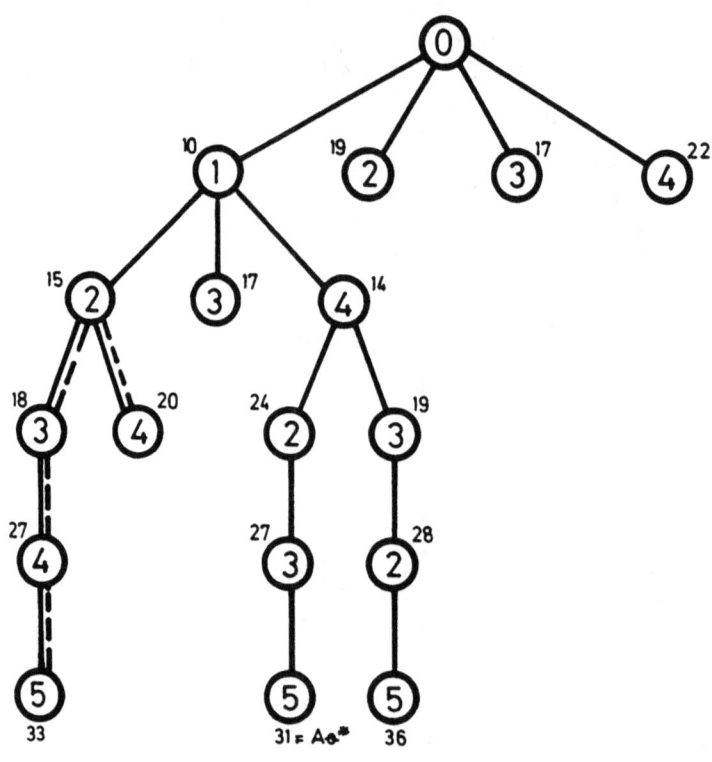

FIGURE 63. Third backtracking.

VEHICLE ROUTING PROBLEMS ON NETWORKS 107

We can continue branching from node 4 since :

$$T_3 = 20 < 31 = A_{\theta*}$$
$$T_3 + B_3 = 23 < 31 = A_{\theta*}$$

We now reach node 3 (Fig. 64).

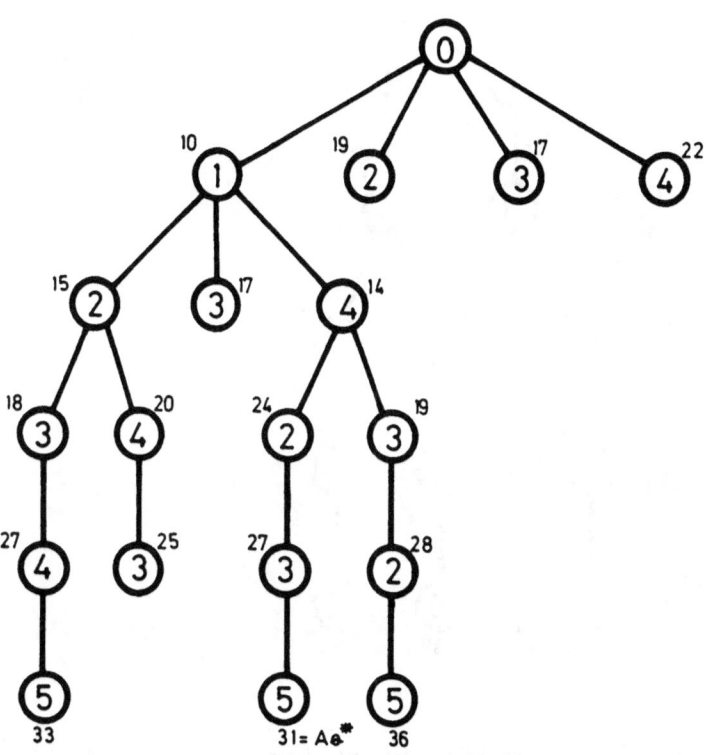

FIGURE 64. Eleventh branching.

We also continue branching from node 3 since :

$$T_4 = 25 < 31 = A_{\theta*}$$
$$T_4 + B_4 = 29 < 31 = A_{\theta*}$$

We have now obtained path $(0, 1, 2, 4, 3, 5)$ (Fig. 65).

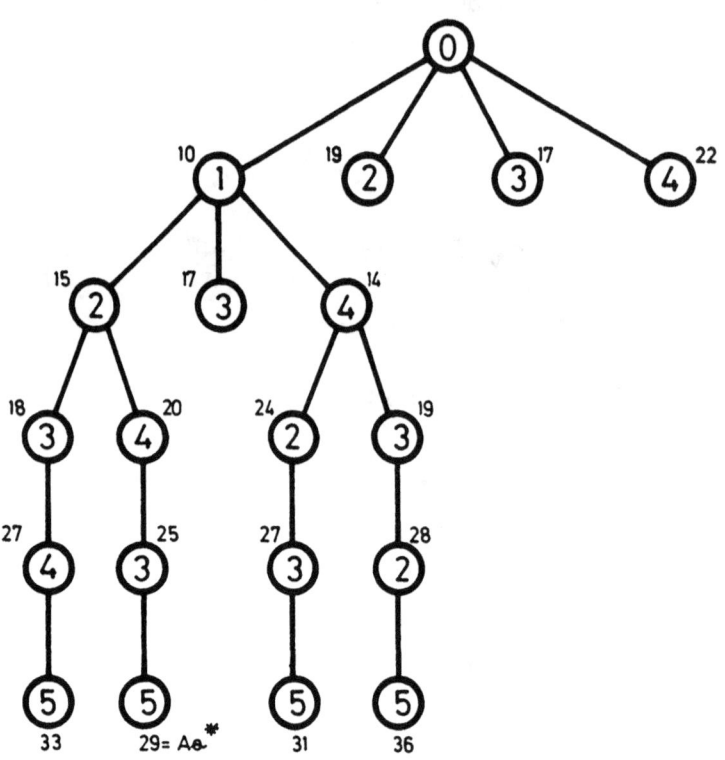

FIGURE 65. Twelfth branching.

Since the new path $\theta = (0, 1, 2, 4, 3, 5)$ fulfills the following :

$$A_\theta = 29 < 31 = A_{\theta *}$$

the upper bound of the optimal solution changes to $A_{\theta *} = 29$.

VEHICLE ROUTING PROBLEMS ON NETWORKS 109

In order to continue branching we return from node 5 to node 3, node 4, node 2, node 1 and up to starting node 0 from which we drop down to the first level of branching in node 3 (Fig. 66).

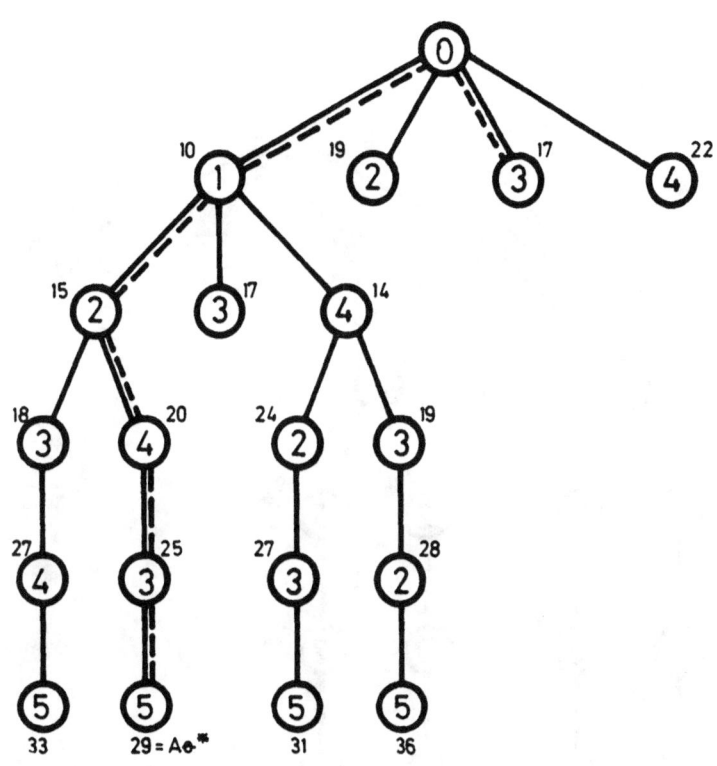

FIGURE 66. Fourth backtracking

For node 3, the first level of branching gives us :

$T_1 = 17 < 29 = A_{\theta *}$

$B_1 = 4 \cdot 4 + 3 \cdot 1 = 19$

$T_1 + B_1 = 36 > 29 = A_{\theta *}$

110 TRANSPORTATION NETWORKS

which means that there is no need to continue branching from this node since the results would be poorer than those we already have.

We return from node 3 to node 0 and then drop down to node 2 on the first branching level (Fig.67).

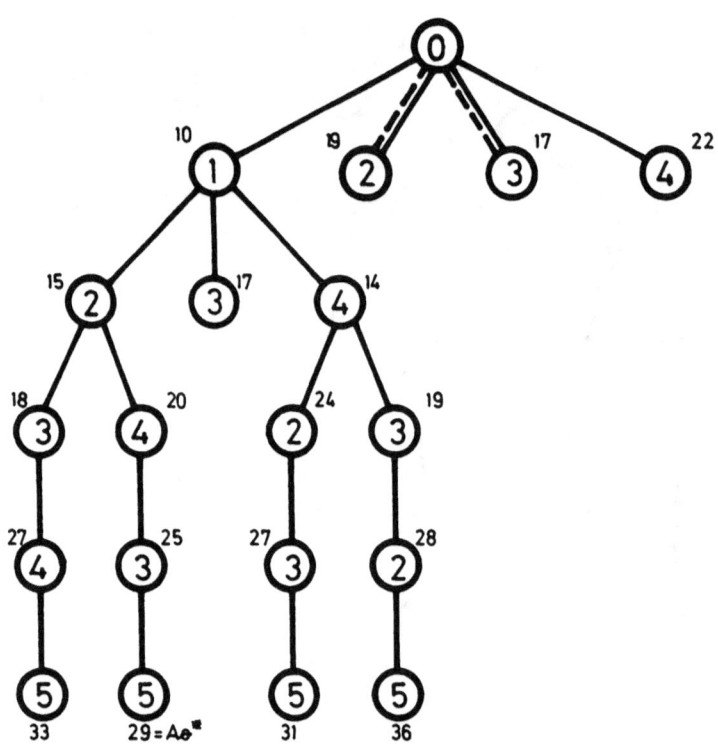

FIGURE 67. Fifth backtracking.

For node 2 we have :

VEHICLE ROUTING PROBLEMS ON NETWORKS 111

$$T_1 = 19 < 29 = A_{\theta*}$$
$$B_1 = 4 \cdot 2 + 3 \cdot 1 = 11$$
$$T_1 + B_1 = 30 > 29 = A_{\theta*}$$

which means that there is no sense in continuing branching from this node either. We return to node 0 and finally go to node 4 on the first level (Fig. 68).

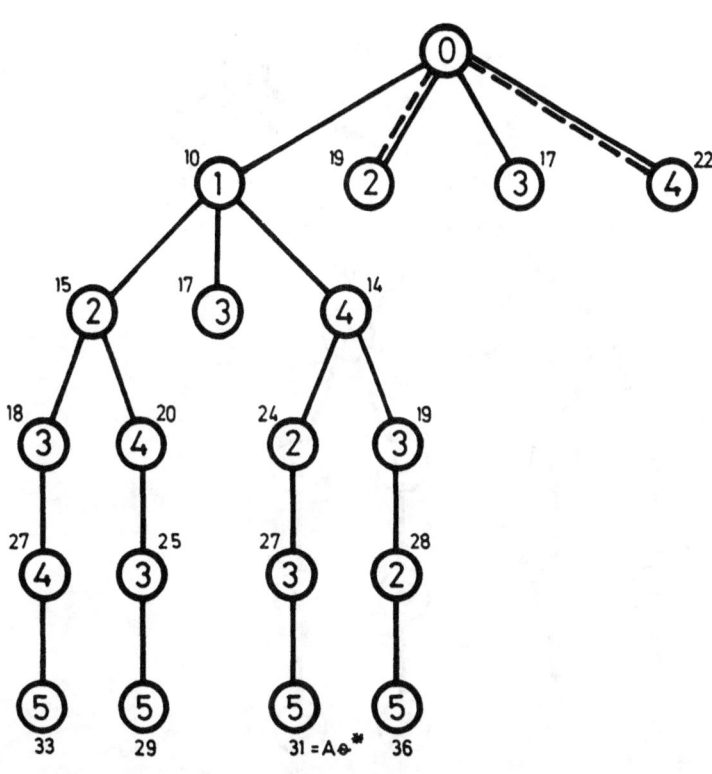

FIGURE 68. Sixth backtracking.

112 TRANSPORTATION NETWORKS

For node 4 we have :
$$T_1 = 22 < 29 = A_{\Theta *}$$
$$B_1 = 4 \cdot 1 + 3 \cdot 1 = 7$$
$$T_1 + B_1 = 29 = 29 = A_{\Theta *}$$

which means we do not continue branching from this node either. Path $\Theta = (0, 1, 2, 4, 3, 5)$ has a total travel time of 29. This is the one and only optimal solution (Fig. 69).

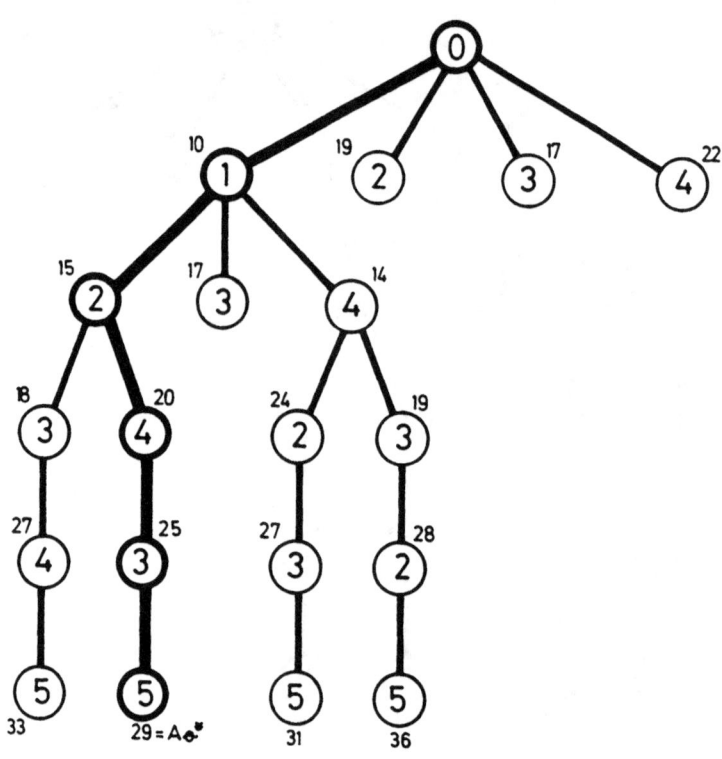

FIGURE 69. Optimal solution.

VEHICLE ROUTING PROBLEMS ON NETWORKS 113

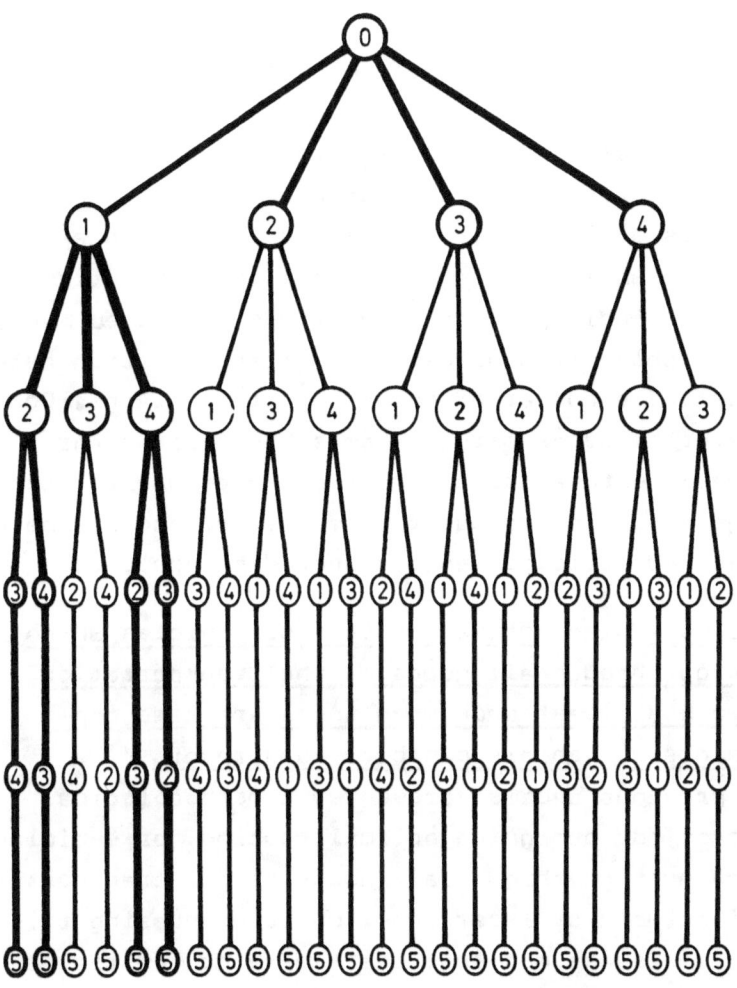

FIGURE 70. Search tree for branch and bound.

114 TRANSPORTATION NETWORKS

The methodological process used to solve this example is known as backtracking. Figure 70 shows the efficiency of this method by a tree containing all possible solutions. The heavy line indicates the branches we went through before finding the optimal solution.

It is clear that solving this problem for a transportation network containing a large number of nodes would require a computer. Experience[2] acquired while solving practical transportation problems indicates that the method described herein is suitable for application on transportation networks whose number of nodes is not too large. Combinatorial programming becomes inefficient for networks with a large number of nodes due to the large amount of computer time needed, and in these cases heuristic procedures should be applied.

3.5. <u>Designing optimal routes for vehicles which must go through all nodes of the transportation network at least once when there are time constraints or vehicle capacity constraints</u>

The previous section presented a methodological process for designing optimal routing for vehicles which must go through all nodes of the transportation network at least once. While discussing this problem we assumed that there were no constraints regarding arrival time at nodes to be serviced or regarding the capacity of the vehicles providing service. We also considered there to be no constraints regarding the maximum utilization of vehicles in terms of authorized work time of those operating the vehicles, etc. Different types of such constraints appear in practical transportation problems depending on the type of transportation and specific problem.

For example, the situation often arises where service in each node of transportation network D_i must be completed by some time d_i. We denote by t_{i_j} the actual time of completing service on node D_i^j which is the j-th station on the route. The following must hold true :

$$t_{i_j} \le d_{i_j} \quad , \quad \forall j$$

Our problem can now be formulated in the following manner : find optimal route $D_0, D_1, \ldots, D_m, D_\emptyset$ for which :

$$A_\Theta = p_0 + \sum_{k=1}^{\emptyset} \left[c(i_{k-1}, i_k) + p_{i_k} \right] \longrightarrow \min$$

and for which:

$$t_{i_j} = p_0 + \sum_{k=1}^{j} \left[c(i_{k-1}, i_k) + p_{i_k} \right] \le d_{i_j}, \forall j$$

Once again $\Theta = (0, i_1, i_2, \ldots, i_m, \emptyset)$ is the permutation of nodes to be visited with total travel time being minimized.

Unless the problem has other requirements, time periods within which service in the starting and finishing nodes must be completed are respectively :

$$d_0 = p_0 \quad d_\emptyset \ge \left[d_i + c(i, \emptyset) \right] \quad 1 \le i \le m$$

This problem can be solved by the same methodological procedure used to solve the example in the previous section. It is clear that existing time constraints for service completion in each node of the transportation network must be taken into consideration when branching and bounding.

In addition to constraints regarding the latest time in which to complete service, constraints often appear regarding the earliest time in which service may begin. Let us consider a situation in which constraints exist concerning the beginning and completion of service in all nodes of the transportation network. We use the following terms:

t_{i_j} – service completion time in node D_i which is the j-th station on the route

s_{i_j} – service starting time in node D_i which is the j-th station on the route

p_{i_j} – duration of service in node D_i which is the j-th station on the route

y_{i_{k-1}, i_k} – time passing between completion of service in node $D_{i_{k-1}}$ and the start **of service in node D_{i_k}**

a_{i_j} – earliest authorized time to start service in node D_i which is the j-th station on the route

d_{i_j} – time limit within which service must be completed in node D_i which is the j-th station on the route.

Our problem now reads:

$$A_\theta = p_0 + \sum_{k=1}^{\emptyset} (y_{i_{k-1}, i_k} + p_{i_k}) \longrightarrow \min$$

$$t_{i_j} = p_0 + \sum_{k=1}^{j} (y_{i_{k-1}, i_k} + p_{i_k}) \leq d_{i_j} \quad \forall_j$$

$$s_{i_j} = t_{i_j} - p_{i_j} \geq a_{i_j} \quad \forall_j$$

For the starting and finishing node we have :

$$a_o = 0 \qquad a_\emptyset = \max_i \left[a_i\right]$$

It should be noted that for problems defined in this manner, the fixed order of visiting nodes $D_o, D_{i_1}, D_{i_2}, \ldots, D_{i_m}, D_\emptyset$ does not at the same time comprise the solution for starting time s_{ij} and completion of service t_{ij} in individual nodes as was the case earlier. To this effect, for any order of visiting nodes e and corresponding value A_e, the value of authorized service completion in some node D_u is found in the interval $t'_u \leqslant t_u \leqslant t''_u$.

We also mentioned that in addition to time constraints, constraints often appear regarding the capacity of vehicles providing service. We denote vehicle capacity (constraints regarding weight, volume or number of seats) by V. We also denote by v_i the amount of goods (number of passengers) to be delivered to node D_i. If :

$$\sum_{i=1}^{m} v_i > V$$

it is clear that one vehicle must start from and return to the depot more than once in order to visit all nodes. In this case, not one but several routes must be formed since the vehicle starts off from the depot several times. We assume that there are no time constraints regarding the start and finish of service in the transportation network nodes. The problem can be finally defined in the following manner :

Determine n consecutive vehicle routes

$$\theta = (o, i_1, i_2, \ldots, i_{j-1}, \emptyset_1, i_{j+1}, \ldots, i_{t-1}, \emptyset_2, i_{t+1}, \ldots i_m, \emptyset_n)$$

with each route starting and finishing in the depot so that

$$A_\theta = \sum_{k=1}^{\emptyset_n} g(i_{k-1}, i_k) \to \min$$

for which :

$$\sum_{i_k=\emptyset_{s-1}}^{\emptyset_s - 1} v_{i_k} \leq V \quad \forall s$$

where $\emptyset_s$ denotes the s-th return to the depot and $g(i_{k-1}, i_k)$ the cost (or time) of transportation from node $D_{i_{k-1}}$ to node D_{i_k}.

We can similarly formulate problems which contain both time or vehicle capacity constraints. For example, constraints often appear regarding the authorized number of nodes to be serviced, the maximum length of one tour, different weight restrictions, etc. Combinatorial programming can be successfully applied to solve these problems. We mention once more, however, that these methods are not very efficient for large transportation networks.

3.6. Heuristic algorithm for the traveling salesman problem

Let us now discuss the following version of the traveling salesman problem :

A vehicle starting and finishing its tour at one fixed point must visit n-1 points. The transportation network connecting these n points is completely connected. This means that it is possible to reach any node from any node, directly, without going through the other nodes (an air transportation network is a typical example of this type of network). The shortest distance between any two nodes equals the length of the branches between these nodes, i.e. if a and b are two of the n nodes under observation, then $d_{ab} = l(a,b)$. From this it ensues that the following inequality is satisfied in our network :

$$l(a,b) \leq l(a,c) + l(c,b)$$

for any three nodes a, b and c.

We also assume that the matrix of shortest distances between the nodes is symmetrical.

The following algorithm is used to solve this simplified problem of the traveling salesman.[29]

Step 1 : Find the shortest spanning tree which connects these n points. We denote this shortest spanning tree by A.

Step 2 : Let k be the number of odd degree nodes of the n nodes (k is always an even number). We pairwise match these k nodes so that the total length of the branches connecting these nodes is minimized. These k nodes with their corresponding branches obtained by pairwise matching comprise a network which we denote by B. We now draw network C which is the union of networks A and B.

Step 3 : Network C does not contain any odd degree nodes. We now draw a Euler tour in network C. This Euler tour is an approximate solution to the traveling salesman problem.

Step 4 : Check which of the nodes is visited more

than once and improve the traveling salesman tour obtained after Step 3 by taking the following inequality into consideration :

$$l(a,b) \leq l(a,c) + l(c,b)$$

It is interesting to establish the relationship between the tour length obtained by the above algorithm with the actual length of the traveling salesman tour. It has been shown (29) that the following holds true :

$$L(C) \leq \frac{3}{2} L(S)$$

where $L(C)$ is the tour length obtained by the heuristic algorithm, and $L(S)$ the actual length of the traveling salesman tour.

E x a m p l e : Find the optimal traveling salesman tour for the network shown in Figure 71 which starts and finishes in node 1.

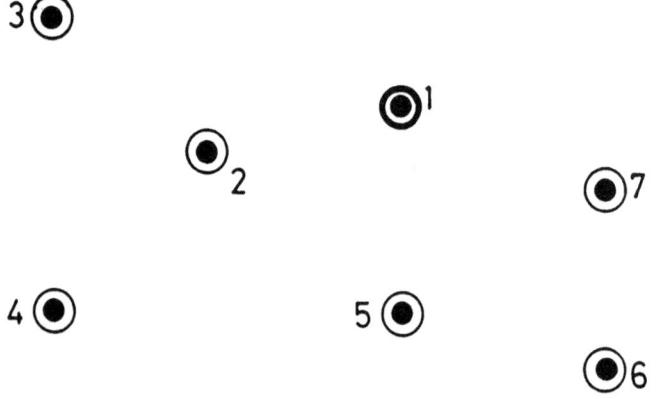

FIGURE 71. Network $G(N,A)$ for the traveling salesman tour.

VEHICLE ROUTING PROBLEMS ON NETWORKS

The distance between pairs of nodes is given in Table V.

TABLE V. Distances between all pairs of nodes.

	1	2	3	4	5	6	7
1	0	75	135	165	135	180	90
2	75	0	90	105	135	210	150
3	135	90	0	150	210	300	210
4	165	105	150	0	135	210	210
5	135	135	210	135	0	90	105
6	180	210	300	210	90	0	120
7	90	150	210	210	105	120	0

We start with Step 1 of the algorithm and find the minimum spanning tree connecting our 7 points. By using the algorithm for finding the minimum spanning tree, we get the tree on Figure 72.

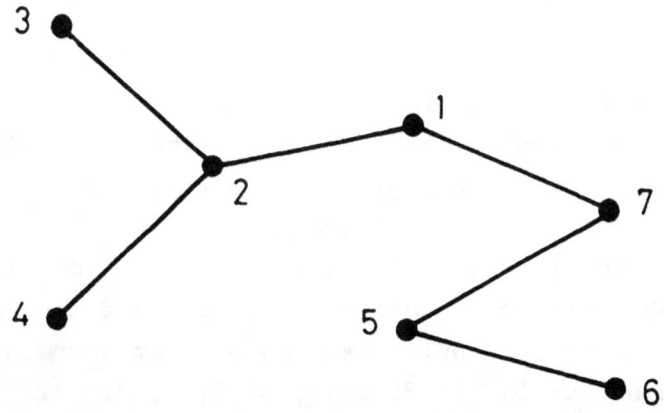

FIGURE 72. Minimum spanning tree A of network $G(N,A)$.

122 TRANSPORTATION NETWORKS

We denote the minimum spanning tree by A. It contains 4 odd degree nodes, nodes 2, 3, 4 and 6.

We move to Step 2 and pairwise match these 4 nodes so that the total branch length connecting them is minimized. The optimal matching is :

2 - 3 and 4 - 6

Nodes 2 and 3, and 4 and 6 with branches (2,3) and (4,6) make up the network we denote by B. This network is shown in Figure 73.

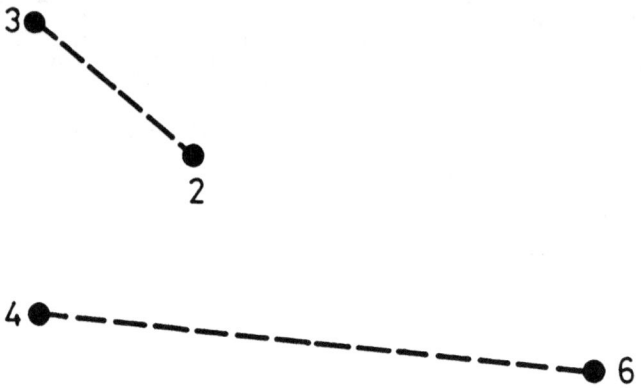

FIGURE 73. Network B.

We now draw network C which is the union of networks A shown on Fig. 72 and B shown on Fig. 73. Network C is shown on Figure 74.

The total length of all branches in network C is 855. Network C does not have any odd degree nodes and we can therefore make a Euler tour in it. One such tour is (1, 2, 3, 2, 4, 6, 5, 7, 1).

We now go to Step 4 and establish which of the nodes in the Euler tour is visited more than once.

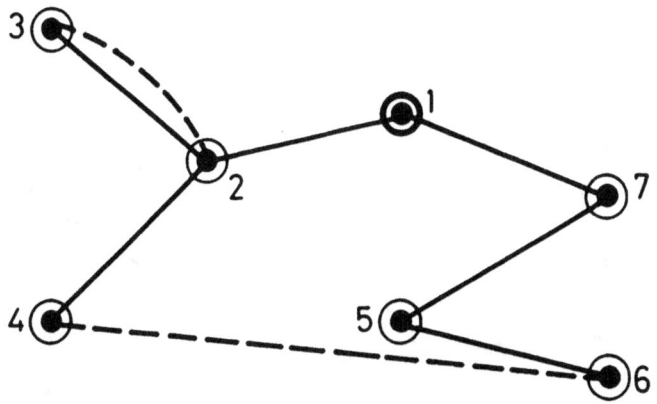

FIGURE 74. Network C.

Node 2 is visited twice. We can replace part of the Euler tour 1, 2, 3, 2, 4... by 1, 3, 2, 4,... or by 1, 2, 3, 4,.... Path 1, 3, 2, 4 is 330 and path 1, 2, 3, 4 is 315. Therefore, we replace part of tour 1, 2, 3, 2, 4,... by 1, 2, 3, 4,... Our final tour reads :
$$(1, 2, 3, 4, 6, 5, 7, 1)$$
This tour with length 810 is shown on Fig. 75.

124 TRANSPORTATION NETWORKS

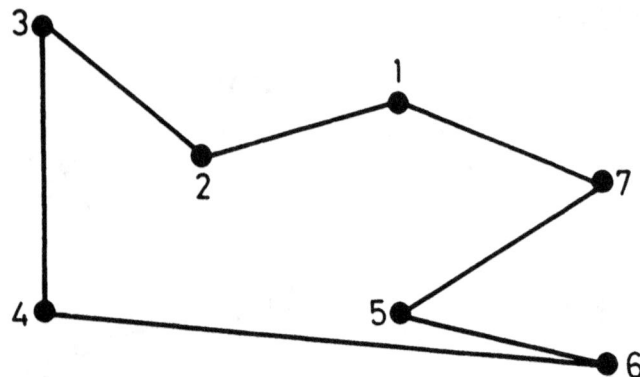

FIGURE 75. Optimal traveling salesman tour.

3.7. Routing and scheduling problems for several vehicles on the transportation network

All problems discussed so far have concerned routine one single vehicle. However, far more important problems concern routing several vehicles on the network. Methods to solve these problems were developed primarily in the past ten years or so, the delay being caused by their extreme complexity which can be successfully solved only by using efficient computer systems.

The algorithms for solving the routing of several vehicles on the network are mainly heuristic. Most often a method of routing one vehicle on the network is used with the fixed geographic region being divided into smaller parts.

To this effect, the fixed region is most often divided into parts and the optimal vehicle tour is found within the framework of the individual parts.

It is also possible to find the optimal tour for the entire region then divide it into parts to be taken by each vehicle.

Both ways are used when solving practical problems. The choice depends on the specific problem to be solved.

3.8. Classification of vehicle routing and scheduling problems on a transportation network

Different versions of vehicle routing and scheduling problems on the transportation network appear in all fields of transportation, depending on the specific problem at hand. Well-organized vehicle routing or a well-designed schedule can markedly contribute towards a decrease in transportation costs and increase the quality of transportation services.

Vehicle routing problems do not have time constraints as to when services in different nodes should start or finish. Contrary to this, scheduling problems contain times fixed in advance within which service in each node must be completed. In cases when a certain time interval is planned for performing services in each node, we usually speak of a combination vehicle routing and scheduling problem. Starting with specific charateristics which describe certain types of routing or scheduling problems, L. Bodin and B. Golden[9] made the following classification :

A. - <u>Time to service in a specific node or on a specific branch</u>
   1. time to carry our service fixed in advance (scheduling problems),

2. service in certain nodes must be carried out within a specific time interval (combined routing and scheduling problems),
3. there are no specific demands regarding service in each node (vehicle routing problems);

B. - Number of vehicle depots in the network
   1. there is only one depot in the network,
   2. the network contains several depots;

C. - Size of vehicle fleet available
   1. the fleet contains only one vehicle,
   2. the fleet contains several vehicles;

D. - Type of vehicles in the fleet
   1. all vehicles in the fleet are the same,
   2. the fleet contains different types of vehicles;

E. Nature of service demands
   1. deterministic demands appear in the network,
   2. stochastic demands for service appear;

F. Location of service demands
   1. service demands appear in the network's nodes,
   2. service demands appear in the network's branches,
   3. service demands appear in both nodes and branches;

G. - Type of transportation network
   1. oriented transportation network,
   2. nonoriented transportation network,
   3. mixed transportation network;

H. - Vehicle capactiy constraints
   1. all vehicles have the same regulated capa-

city constraints,
2. there are differences between vehicles regarding regulated capacity constraints,
3. there are no constraints regarding vehicle capacity;

I. - <u>Maximum allowed vehicle route length</u>
1. all vehicles in the fleet have the same maximum allowed route length,
2. some vehicles have different maximum allowed route lengths,
3. there are no constraints regarding the maximum allowed vehicle route length;

J. - <u>Costs</u>
1. variable,
2. fixed;

K. - <u>Operations carried out</u>
1. picking up,
2. delivering,
3. picking up and delivering (loading and unloading goods or picking up and dropping off passengers);

L. - <u>Objective functions on which optimization is based</u>
1. minimizing route costs,
2. minimizing total fixed and variable costs,
3. minimizing the number of vehicles needed to carry out transportation operations;

M. - <u>Other constraints</u> (depends on specific problem)

The above classification is very useful, particularly because it gives a better insight into specific transportation problems to be solved. It also helps when making analogies between problems in different branches of transportation and solving them by the same or similar methodolotical process.

3.9. Designing optimal routes for a fleet of vehicles which must service every node on the transportation network

Let us now consider the case when m traveling salesmen must visit n points with each of the m tours starting and finishing at the same point. We denote this point by B. In this case we must find m minimum-length routes which do not have any common branches, which "cover" all the points to be visited and which all start and finish at point B. We now replace this point B by m copies $B_1, B_2, \ldots, B_m$ with each of these copy-points being connected to the set of n points to be visited in the same way that original point B is connected to them. Let the distances between copy-points and the points in the set of n points be equal to the distance between original point B and individual points from the set of n points to be visited, i.e. let

$$d(B_1, a) = d(B_2, a) = \ldots = d(B_m, a) = d(B, a)$$

where a is one of the nodes to be visited.

Let the distances between individual copy-points be infinitely large

$$d(B_i, B_j) = \infty \quad za \quad \forall_{i,j} = 1, 2, \ldots, m$$

The problem can now be treated as a classical traveling salesman problem, i.e. as a problem in which one traveling salesman must visit (m + n) points exactly once. When solving this problem, node B_i will not be followed by node B_j since

$d(B_i, B_j) = \infty$ za $\forall_{i,j} = 1, 2, \ldots, m.$

When solving the classical traveling salesman problem, all copies must be joined to original point B and the single traveling salesman tour will be decomposed into m smaller tours as required in the original problem.

E x a m p l e : Figure 76 shows a transportation network which contains a depot and 4 nodes to be visited by 2 vehicles. The total length of both tours should be minimized. Fig. 76 also indicates the length of individual branches.

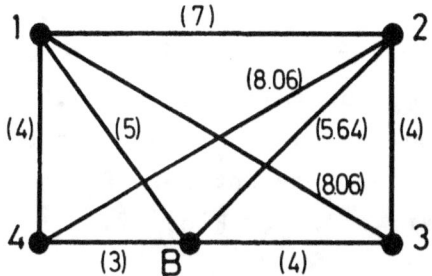

FIGURE 76. Network with one depot and 4 nodes to be visited.

The matrix of distances between nodes is symmetrical :

	B	1	2	3	4
B	∞	5	5,64	4	3
1	5	∞	7	8,06	4
2	5,64	7	∞	4	8,06
3	4	8,06	4	∞	7
4	3	4	8,06	7	∞

Since there are 2 vehicles, 2 copies of depot B must be made. We denote them by B_1 and B_2. The new distance matrix is now:

$$\begin{array}{c} \\ B_1 \\ B_2 \\ 1 \\ 2 \\ 3 \\ 4 \end{array} \begin{array}{c} \begin{array}{cccccc} B_1 & B_2 & 1 & 2 & 3 & 4 \end{array} \\ \left[\begin{array}{cccccc} \infty & \infty & 5 & 5{,}64 & 4 & 3 \\ \infty & \infty & 5 & 5{,}64 & 4 & 3 \\ 5 & 5 & \infty & 7 & 8{,}06 & 4 \\ 5{,}64 & 5{,}64 & 7 & \infty & 4 & 8{,}06 \\ 4 & 4 & 8{,}06 & 4 & \infty & 7 \\ 3 & 3 & 4 & 8{,}06 & 7 & \infty \end{array} \right] \end{array}$$

By applying the algorithm for solving the traveling salesman problem we get the optimal solution:

$$B_1 - 2 - 3 - B_2 - 4 - 1 - 4 - B_1$$

The total length of this tour is 27.64.

By joining copies B_1 and B_2 into one point (point B), 2 tours are obtained for the 2 vehicles. These tours are shown on Fig. 77.

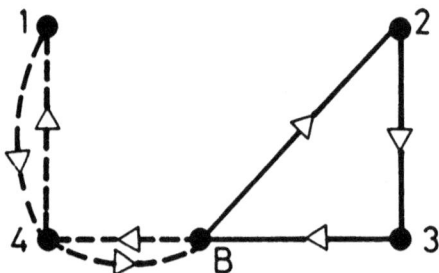

FIGURE 77. Two vehicle tours.

Node 4 is visited twice. By applying the 4th step of the heuristic algorithm for solving the traveling salesman problem, a shorter tour is obtained which reads :

$$B_1 - 2 - 3 - B_2 - 4 - 1 - B_1$$

This tour has a length of 25.64. The optimal vehicle routes are shown on Fig. 78.

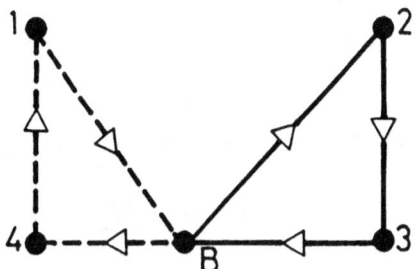

FIGURE 78. Optimal vehicle tours.

3.10. The problem of routing several vehicles when there is only one depot

Now, within the field under discussion, let there be n nodes to service, each demanding v_i (i = 1, 2,...,n) amount of some good, or in the other case each demanding the transportation of v_i passengers. Vehicles are stationed at point B.

All vehicles have the same capacity V and when servicing all must start and finish their trips at point B. Let capacity V of the vehicles be greater than any amount of goods v_i in demand or the number of passengers to be transported so that each point is serviced by only one vehicle or one vehicle can service several points.

The problem presented to us is that of determining sets of routes to be used by the vehicles when in service so that the total distance covered by the entire fleet of vehicles is at a minimum.

One of the best algorithms for solving this problem is Clark-Wright's "saving" algorithm,[13] which is actually very simple.

We now study point B and the n points to be serviced. Let one vehicle service **one** point at the beginning. This means at the beginning n vehicles leave point B, service n points and return to point B. The total distance covered by all n vehicles is:

$$2d(B,1) + 2d(B,2) + 2d(B,3) + \ldots + 2d(b,n) = 2\sum_{i=1}^{n} d(B,i)$$

where $d(B,i)$, $(i = 1, 2, \ldots, n)$ is the distance between point B and point i.

If one vehicle should service two points instead of one, let's say i and j, then there is a saving made which equals :

$$S(i,j) = 2d(B,i) + 2d(B,j) - \left[d(B,i) + d(i,j) + d(B,j)\right] = d(B,i) + d(B,j) - d(i,j)$$

Quantity $S(i,j)$ is called the "saving" which is obtained by joining points i and j into one route. It is clear that the larger $S(i,j)$ becomes, the better it is to join i and j into one route. Points i and j cannot be joined into one trip if doing so violates one of the problem's constraints (constraints regarding vehicle capacity, maximum allowed route length, number of nodes which can be reached, etc.).

Clark-Wright's "saving" algorithm is composed of the following algorithmic steps :

Step 1 : Calculate saving $S(i,j) = d(B,i) + d(B,j) - d(i,j)$ for every pair (i,j) of points to be serviced.

Step 2 : Rank all the savings and compare them by size. Make a list of savings starting with the biggest saving.

Step 3 : When examining savings $S(i,j)$, corresponding branch (i,j) is included in the route, if doing so does not violate one of the given constraints and if :
   (a) neither point i nor point j has been included in a route,
   (b) either point i or point j is already included in a route if that point is not an internal point on the route (an internal route point is not next to starting point B),
   (c) both points i and j are included in different routes and neither one is an internal route point (both are external) in which case the routes can be joined together.

Step 4 : If the list of savings (after formation of the first route) is not completely used up, return to Step 3 and start from the beginning with the largest unused saving. When the list is used up then the algorithm is finished since all routes have been formed.

E x a m p l e : The vehicle depot is located at point 1. The points to be serviced are denoted by 2, 3, 4,..., 9, respectively (Fig. 79).

134 TRANSPORTATION NETWORKS

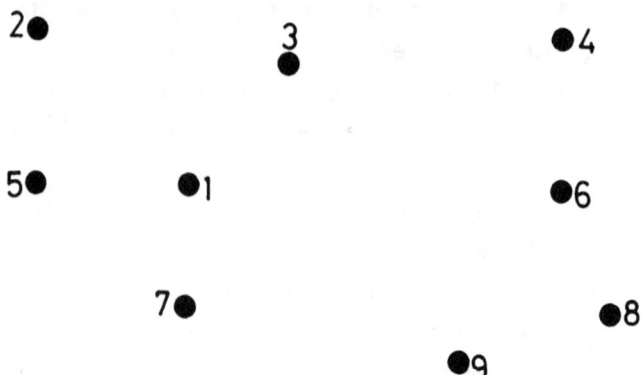

FIGURE 79. Depot 1 and nodes to be serviced.

The distance between individual points is given in Table VI :

TABLE VI. Distance between points.

	1	2	3	4	5	6	7	8	9
1	∞	40	30	58	32	52	28	67	55
2	40	∞	43	81	29	87	63	106	94
3	30	43	∞	37	53	46	57	70	70
4	58	81	37	∞	88	27	77	57	67
5	32	29	53	88	∞	85	40	96	80
6	52	87	46	27	85	∞	62	30	43
7	28	63	57	77	40	62	∞	62	40
8	67	106	70	57	96	30	62	∞	25
9	55	94	70	67	80	43	40	25	∞

The vehicles servicing these points have a capacity of $V = 12$. The quantity of goods (number of passengers) to be picked up from points

2, 3,..., 9 are given in Table VII.

TABLE VII Quantities to be picked up.

Node i	2	3	4	5	6	7	8	9
Quantity v_i	4	7	3	2	6	3	2	3

Using Clark-Wright's algorithm, design the service routes for the vehicles to use.

We calculate the first saving using the formula:

$$S(i,j) = d(1,i) + d(1,j) - d(i,j)$$

Then, for example :

$S(4,6) = d(1,4) + d(1,6) - d(4,6) = 58 + 52 - 27 = 83$

Corresponding savings are calculated for all pairs of nodes. These savings, compared by size, are shown in Table VIII.

Branch (8,9) has the greatest saving. Therefore, our first route for now is $(1, 8, 9, 1)$.

The quantity of goods in the vehicle will be :

$$v_8 + v_9 = 2 + 3 = 5 < 12 = V$$

As can be seen, we are able to join nodes 8 and 9 since this does not violate any constraint concerning vehicle capacity size.

The second by order of savings is branch (6,8). Point 6 could be included in the route since point 8 is not an internal point (point 8 is next to starting point 1). We now check to see whether including point 6 into the route will violate the capacity constraint. So we have :

$$v_8 + v_9 + v_6 = 2 + 3 + 6 = 11 < 12 = V$$

and we conclude that point 6 can be included in the route, so that our route is changed to :

$$(1, 6, 8, 9, 1).$$

TABLE VIII. Calculated savings.

Branch (i,j)	Saving S(i,j)
(8,9)	97
(6,8)	89
(4,6)	83
(4,8)	68
(6,9)	64
(3,4)	51
(4,9)	46
(7,9)	43
(2,5)	43
(3,6)	36
(7,8)	33
(3,8)	27
(2,3)	27
(5,7)	20
(6,7)	18
(2,4)	17
(3,9)	15
(4,7)	9
(3,5)	9
(5,9)	7
(2,6)	5
(2,7)	5
(5,8)	3
(4,5)	2
(3,7)	1
(2,8)	1
(2,9)	1
(5,6)	− 1

Branch (4,6) is the next by order of savings. Point 6 is not an internal point so there is no constraint in this regard to including point 4 in the route. However, since :
$$v_8 + v_9 + v_6 + v_4 = 2 + 3 + 6 + 3 = 14 > 12 = V$$
point 4 cannot be included in the route since this would violate the capacity constraint. Branch (4,8) cannot join the route since point 8 is an internal point and the capacity constraint would be violated as well. Branch (6,9) is also omitted since points 6 and 9 are already included in the route. The next branch on the saving list is branch (3,4). Since neither point 3 nor point 4 is included in the existing route, we open a new route as follows :
$$(1, 3, 4, 1).$$
The amount of goods in the second vehicle equals :
$$v_3 + v_4 = 7 + 3 = 10 < 12 = V$$
Concerning branch (4,9), we note that points 4 and 9 belong to different routes and both are external points. However, these two routes cannot be joined since this would violate the vehicle capacity constraint. The next branch is (7,9). Point 9 is an external point in the first route, but point 7 cannot join this route due to capacity constraints. Now we come to branch (2,5). Since points 2 and 5 do not belong to the two previous routes, we open a third route as follows :
$$(1, 2, 5, 1).$$
This route has capacity :
$$v_2 + v_5 = 4 + 2 = 6 < 12 = V$$
Now we examine branch (3,6). Points 3 and 6 are external points in two different routes. How-

ever, we cannot join these two routes due to capacity constraints. The next branch is (7,8). Point 7 cannot join the first route because point 8 is an internal point of this route and also because of capacity constraints. In the case of branch (3,8), we note that these points belong to different routes which cannot be joined. This also holds true for branch (2,3). Branch (5,7) appears next. Point 5 is an external point in the third route. Also :

$$v_2 + v_5 + v_7 = 4 + 2 + 3 = 9 < 12 = V$$

We conclude that point 7 can be included in the third route which now contains :

$$(1, 2, 5, 7, 1).$$

The first, second and third routes are shown on Figure 80.

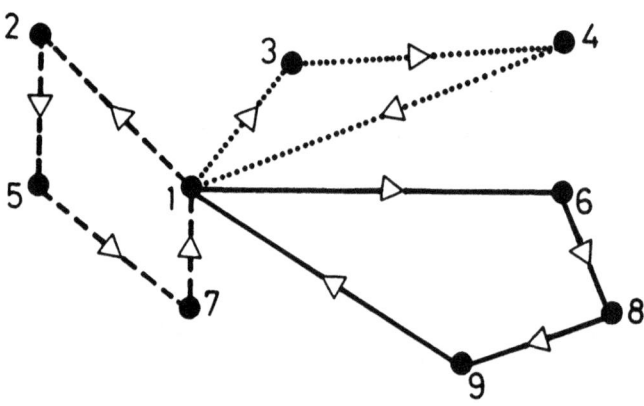

FIGURE 80. Three vehicles routes.
All three routes have a total distance of 424.

3.11. Heuristic "sweeping" algorithm to route vehicles on a transportation network when there is one depot

Professional literature contains several modifications of the Clark-Wright algorithm for routing vehicles when there is only one depot. One of the best-known is the "sweeping" algorithm which was given by B. Gilett and L. Miller in 1974.[24]

This algorithm is applied to polar coordinates and the depot is considered to be the origin of the coordinate system. Then the depot is joined with an arbitrarily chosen point which is called the seed point. All other points are joined to the depot and then aligned by increasing angles which are formed by the segments which connect the points to the depot and the segment which connects the depot to the seed point. The route starts with the seed point and then the points aligned by increasing angles are included, respecting given constraints all the while. When a point cannot be included in the route since this would violate a given constraint, this point becomes the seed point of a new route, etc. The process is completed when all points are included in a route.

Let us return to the example used to show the efficiency of the Clark-Wright algorithm. Point 1 containing the depot is now the system origin and point 3 is the seed point. In this case, we measure the angles in clockwise direction (the angle measurement direction is chosen arbitrarily).

The origin of the system, seed point, other points and corresponding angles are shown on Fig. 81.

140 TRANSPORTATION NETWORKS

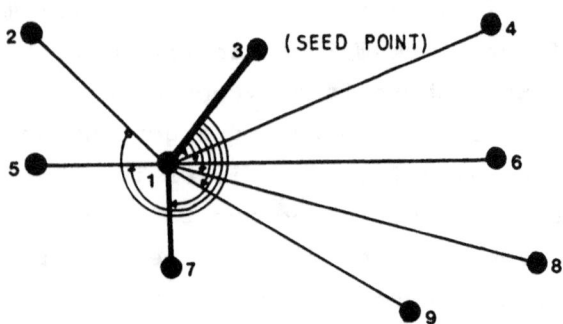

FIGURE 81. Sweeping algorithm after first step (seed point).

The route starts from the depot (point 1) and goes to the seed point (point 3). From point 3 should be taken 7 units of goods (or 7 passengers). The route now contains (1, 3). We now include the next point to the route from the set of points aligned in increasing angles. This is point 4 which has 3 units to be transported which makes a total of 10 units when added to those of point 3. Since this is less than the vehicle capacity of 12 units, point 4 can be included in the route. The new route now contains (1, 3, 4). Point 6 cannot be included in the route since the route (1, 3, 4, 6) would have to carry 16 units which is in excess of the vehicle capacity. So we have finished with the first route structure which reads (1, 3, 4, 1). Point 6 becomes the new seed point. Figure 82 shows

the first route (1, 3, 4, 1) and the new seed
point - point 6, with the other points and corresponding angles.

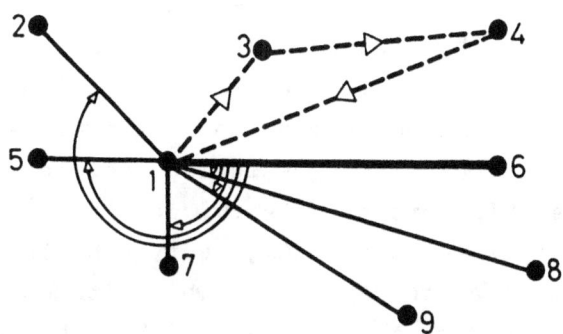

FIGURE 82. Sweeping algorithm after second
step (first route and new seed point).

By continuing with the described procedure,
we get two more routes - route (1, 6, 8, 9, 1)
and route (1, 7, 5, 2, 1) which is the solution
shown in Fig. 83.

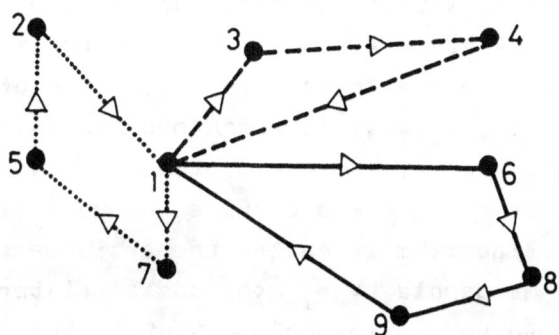

FIGURE 83. Sweeping algorithm - final routing.

3.12. Heuristic "assignment" algorithm to route vehicles on a transportation network when there is one depot

One of the best modifications of Clark-Wright's algorithm to rotate vehicles when there is only one base is the Fisher-Jaikumar "assignment" algorithm which dates from 1981.[17] The Fisher-Jaikumar algorithm is made up of two parts. The first part consists of assigning vehicles to groups of nodes which are to be serviced. At the end of this part of the algorithm, every vehicle is assigned to a group of nodes which it must service. In the second part of the algorithm, the route length of vehicles servicing groups of nodes is optimized. This optimization is made using some of the algorithms for solving the traveling salesman problem.

Let there be a total K vehicles in service. We denote them respectively by 1, 2, 3,..., K. The number of nodes to be serviced is denoted by n, and c_{ij} denotes the costs (distance, time) involved in traveling from node i to node j. In the Fisher-Jaikumar algorithm of assignment, first K nodes are arbitrarily chosen from the set of n nodes which are to be serviced. We denote these nodes by $i_1, i_2,..., i_k$. Each node is joined to one vehicle 1, 2,..., K. In other words, each vehicle is assigned to service one node (which is why the algorithm is called the assignment algorithm). We denote by d_{ik} the costs (distance, time) involved in inserting node i into the route segment which vehicle k follows, starting from the depot (**node 0**), servicing node i_k and returning to the depot (node 0). Node i will be serviced by vehicle K as follows :

$$d_{ik} = \min \left[c_{oi} + c_{ii_k} + c_{i_k o}; c_{oi_k} + c_{i_k i} + c_{io} \right] - \left[c_{oi_k} + c_{i_k o} \right]$$

There is a computer program developed to choose the node of origin. However, these nodes should be chosen without a computer at first since whoever is solving the problem has a sort of built-in feeling as to which node should be the node of origin. Nodes of origin are usually chosen from those nodes which are located farthest from the vehicle depot. Nodes of origin are often chosen for which :

$$a_i > \frac{1}{2} b_k$$

where a_i is the quantity of goods (number of passengers) to be delivered (picked up) at node i, and b_k is the capacity of vehicle k.

Node i for which $a_i > \frac{1}{2} b_k$ is often chosen as the node of origin since no pairs of these nodes can be on the same route for this would violate the vehicle capacity constraint.

E x a m p l e : Let us now show the possibility of applying the Fisher-Jaikumar algorithm on the same example we used for the Clark-Wright and the Gillett-Miller heuristic sweeping algorithm. The vehicle depot is located at point 1. The points to be serviced are denoted by 2, 3, 4,..., 9, respectively (Fig. 84).

The distances between individual points are given in Table IX.

144 TRANSPORTATION NETWORKS

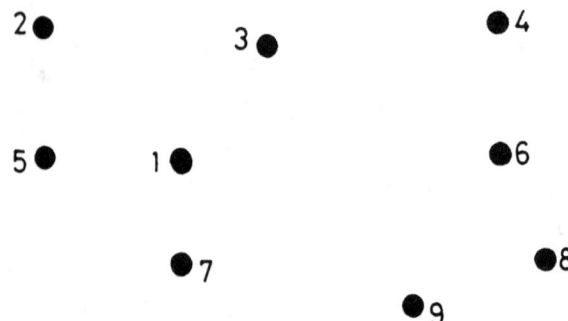

FIGURE 84. Depot 1 and points to be serviced.

TABLE IX. Distances between points.

	1	2	3	4	5	6	7	8	9
1	∞	40	30	58	32	52	28	67	55
2	40	∞	43	81	29	87	63	106	94
3	30	43	∞	37	53	46	57	70	70
4	58	81	37	∞	88	27	77	57	67
5	32	29	53	88	∞	85	40	96	80
6	52	87	46	27	85	∞	62	30	43
7	28	63	57	77	40	62	∞	62	40
8	67	106	70	57	96	30	62	∞	25
9	55	94	70	67	80	43	40	25	∞

The capacity of the servicing vehicle is $V = 12$. The quantity of goods (number of passengers) to be picked up from points 2, 3,..., 9 are given in Table X.

VEHICLE ROUTING PROBLEMS ON NETWORKS 145

TABLE X. Quantity of goods to be picked up.

Node i	2	3	4	5	6	7	8	9
Quantity v_i	4	7	3	2	6	3	2	3

Let there be 3 vehicles in service. We denote them respectively by 1, 2 and 3. We arbitrarily choose 3 points as the system origins. Let the nodes be 2, 4 and 9. The three routes which go from the depot to these nodes and back again are shown in Fig. 85.

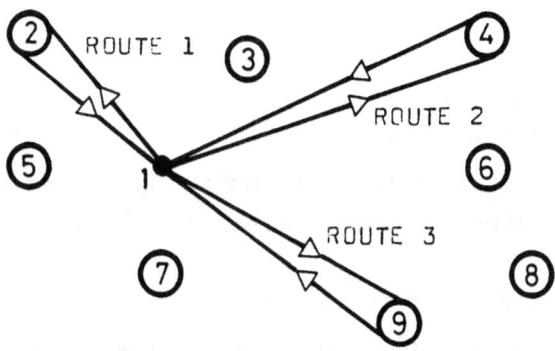

FIGURE 85. Three initial routes.

We start with node 3. If we include it in the first route ($k = 1$) the length of the first route is increased by :

$$d_{31} = (40 + 43 + 30) - (40 + 40) = 33$$

If we include node 3 in the first or second route the lenghts of these routes are increased by:

146 TRANSPORTATION NETWORKS

$$d_{32} = (58 + 37 + 30) - (58 + 58) = 9$$
$$d_{33} = (55 + 70 + 30) - (55 + 55) = 45$$

We include node 3 in route 2 which now contains $(1, 3, 4, 1)$. The vehicle on this route segment will transport $v_3 + v_4 = 7 + 3 = 10$ units of goods (passengers). Since $10 < 12 = V$ we are permitted to join nodes 3 and 4 into one route.

We calculate the route increase for nodes 5, 6 and 8 as follows :
$$d_{51} = (40 + 29 + 32) - (40 + 40) = 21$$
$$d_{52} = (32 + 53 + 37 + 58) - (30 + 37 + 58) = 55$$
$$d_{53} = (32 + 80 + 55) - (55 + 55) = 57$$

We include node 5 into route 1 which now contains $(1, 5, 2, 1)$. The vehicle must transport $v_5 + v_2 = 2 + 4 = 6$ on this route segment.

Since $6 < 12 = V$ this route construction is also permitted. For node 6 we have :
$$d_{61} = (52 + 85 + 29 + 40) - (32 + 29 + 40) = 105$$

We do not calculate quantity d_{62} since adding node 6 to route 2 would exceed the vehicle capacity limit.
$$d_{63} = (52 + 43 + 55) - (55 + 55) = 40$$

We include node 6 into route 3 which now contains $(1, 6, 9, 1)$. This route segment will transport $v_6 + v_9 = 6 + 3 = 9$ units of goods. This route construction is permitted since $9 < 12 = V$.

Nodes 8 and 7 are still not assigned. Then :
$$d_{81} = (67 + 96 + 29 + 40) - (32 + 29 + 40) = 131$$
$$d_{82} = (67 + 70 + 37 + 58) - (30 + 37 + 58) = 107$$
$$d_{83} = (67 + 30 + 43 + 55) - (52 + 43 + 55) = 45$$

We include node 8 into route 3 which contains

(1, 8, 6, 9, 1). This route will transport $v_8 + v_6 + v_9 = 2 + 6 + 3 = 11$ units of goods.

Node 7 can only be included in route 1, for if it is added to either route 2 or route 3 the vehicle capacity constraint is exceeded. Route 1 contains (1, 7, 5, 2, 1). Routes 1, 2 and 3 are shown in Fig. 86.

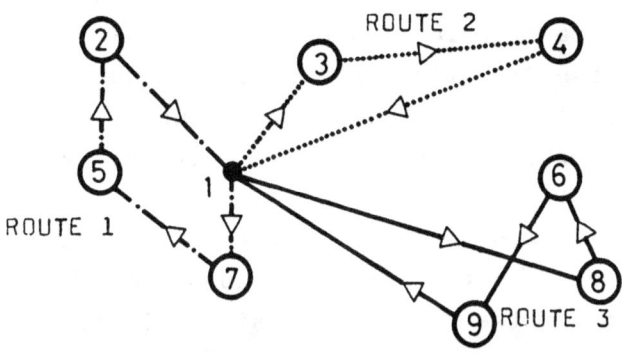

FIGURE 86. Three routes obtained after the first part of the Fisher-Jaikumar algorithm.

The first part of the Fisher-Jaikumar algorithm is finished since all three vehicles are assigned to service different groups of nodes. The second part of the algorithm is to optimize the length of different routes. As already mentioned, this algorithmic step uses some of the algorithms for solving the traveling salesman problem. The very small transportation network of our example enables us to easily determine the optimal routes. They are :

Route 1 : (1, 7, 5, 2, 1)   route length is 137
Route 2 : (1, 3, 4, 1)   route length is 125
Route 3 : (1, 6, 8, 9, 1)   route length is 162

Figure 87 shows routes 1, 2 and 3 in the solution obtained using the Fisher-Jaikumar algorithm. For purposes of comparison, the same figure shows the solutions obtained earlier using Gillett-Miller's and Clark-Wright's algorithms.

By applying the three algorithms, we obtained three solutions with an equal total route length. The Fisher-Jaikumar and Gillett-Miller algorithms gave the same solution. However, it should be noted that different solutions are most often obtained due to the algorithms' heuristic nature. The difference between the solutions obtained by applying different algorithms are not very significant, although it has been shown that the best solutions are obtained from the Fisher-Jaikumar algorithm.

By varying the number of vehicles in service, one can determine the smallest number of vehicles which is able to provide service to all nodes on the transportation network. The Fisher-Jaikumar algorithm can be used to design optimal routes for a given number of vehicles and to determine the minimum number of vehicles for a given set of nodes to be serviced.

VEHICLE ROUTING PROBLEMS ON NETWORKS 149

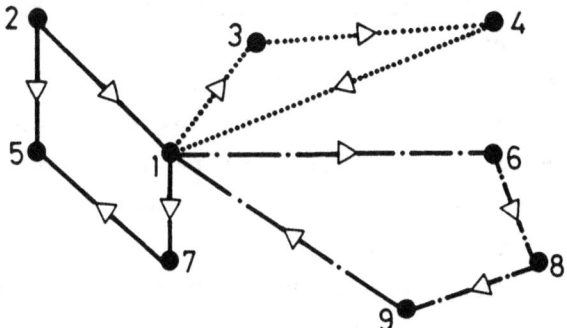

Fisher-Jaikumar algorithm solution - length 424.

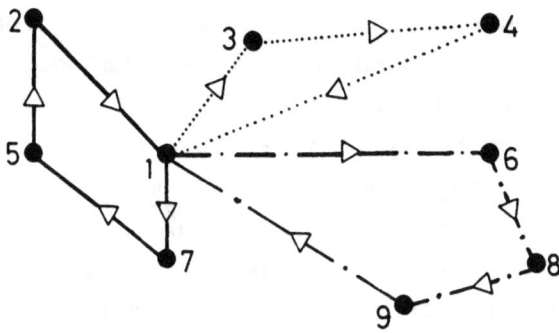

Gillett-Miller algorithm solution - length 424.

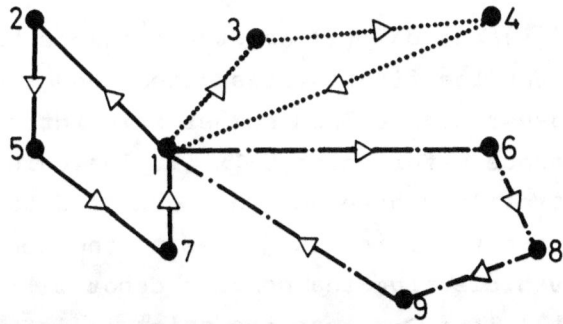

Clark-Wright algorithm solution - length 424.

FIGURE 87. Comparison of the three solutions.

3.13. The problem of vehicle routing when there are several depots

The problem of routing vehicles when there are several depots is even more complex than the same problem with one depot. When there are several depots the problem appears of joining points which are serviced by individual depots. The problem of vehicle routing on a network with several depots is most often solved in two steps. In the first step, individual depots are joined to groups of points to be serviced. The second step solves the problem of routing vehicles between each depot and its corresponding group of points, using the methods described earlier.

One of the best-known methods for joining individual depots to groups of points was developed in 1974 by B. Gillett and J. Johnson.[23] The method is as follows : first the following relation is calculated for each point i to be serviced :

$$a_i = \frac{d^{(1)}(i)}{d^{(2)}(i)}$$

where $d^{(1)}(i)$ and $d^{(2)}(i)$ are the distance between point i and the first closest depot and the second closest depot. Then number x is introduced into the process for which $0 < x < 1$. The value of x is arbitrarily chosen and then compared to a_i. If $a_i \leq x$ then the point is joined to the nearest depot (a vehicle from the nearest depot will service it). If $a_i > x$ then the point is left for further consideration. When all the points for which $a_i \leq x$ are joined to corresponding depots, the points for which $a_i > x$ are taken into consideration. These points are joined to depots as follows. Let there be two points b and c joined to a depot B_p. If we want to add point a to the route between b and c and join it to depot B_p then we increase the route length starting from depot B_p by :

$$d_{bc}(a) = d_{ba} + d_{ac} - d_{bc}$$

It is clear that we will joint point a to the depot where its addition will cause the least increase in the route length starting from this depot. When all points have been joined to depots in this manner, the algorithms developed for the case of one depot are applied.

E x a m p l e : Figure 88 shows 12 nodes. Vehicle depots are located at nodes 1, 2 and 3. Applying the Gillett-Johnson algorithm, join depots to groups of nodes they will service.

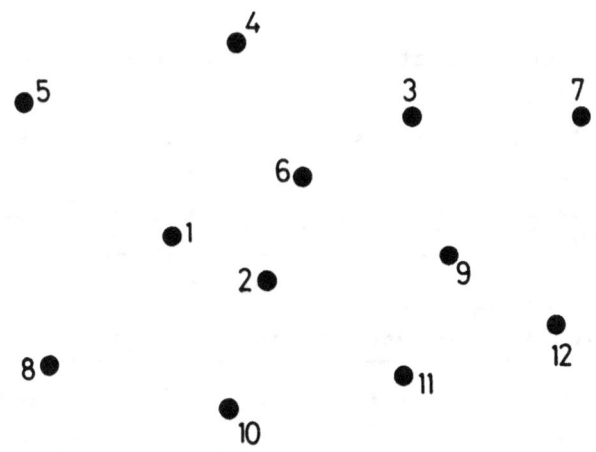

FIGURE 88. Depots 1, 2 and 3 and service points.

152 TRANSPORTATION NETWORKS

Distances between individual nodes are given in the following matrix :

	1	2	3	4	5	6	7	8	9	10	11	12
1	–	28	70	32	41	45	120	36	75	35	60	103
2	28	–	55	47	70	35	100	53	51	28	32	76
3	70	55	–	60	100	22	54	102	33	83	53	56
4	32	47	60	–	44	40	112	63	80	65	75	107
5	41	70	100	44	–	80	154	44	113	69	100	142
6	45	35	22	40	80	–	75	80	40	63	65	67
7	120	100	54	112	154	75	–	153	53	127	83	46
8	36	53	102	63	44	80	153	–	104	35	80	128
9	75	51	33	80	113	40	53	104	–	75	30	28
10	35	28	83	65	69	63	127	35	75	–	47	95
11	60	32	53	75	100	65	83	80	30	47	–	49
12	103	76	56	107	142	67	46	128	28	95	49	–

For every node $i = 4, 5, \ldots, 12$, calculate distances $d^{(1)}(i)$ and $d^{(2)}(i)$ to the first and second closest depots. These distances are given in Table XI.

TABLE XI. Distances to first and second closest depots.

Node i	4	5	6	7	8	9	10	11	12
$d^{(1)}(i)$	32	41	22	54	36	33	28	32	56
$d^{(2)}(i)$	47	70	35	100	53	51	35	53	76

For every node $i = 4, 5, \ldots, 12$ calculate the relation :

$$a_i = \frac{d^{(1)}(i)}{d^{(2)}(i)}$$

VEHICLE ROUTING PROBLEMS ON NETWORKS 153

The numbers for a_i are given in Table XII.

TABLE XII. Values for a_i.

Node i	4	5	6	7	8	9	10	11	12
a_i	0,68	0,59	0,63	0,54	0,70	0,65	0,8	0.60	0,74

Arbitrarily choose a value for x. Let

$x = 0.75$

All nodes i for which $a_i \leq 0.75$ are joined to the nearest depot. This is shown in Table XIII.

TABLE XIII. Joining nodes to nearest depots.

Node i	Joined to depot:
4	1
5	1
6	3
7	3
8	1
9	3
11	2
12	3

Since :
$$a_{10} = 0,8 > 0,7 = x$$
node 10 is not joined to a depot for the moment.

We now consider what would happen if node 10 was joined to depot 1. Nodes 4, 5 and 8 are already joined to depot 1. If node 10 is added to the route between points 4 and 5, the route length starting from depot 1 will be increased by :

$d_{4,5}(10) = d_{4,10} + d_{10,5} - d_{4,5} = 65 + 69 - 44 = 90$

Then we calculate how much the route would be increased from depot 1 if we added node 10 between nodes 4 and 8. This is:

$$d_{48}(10) = 37$$

For the pair of nodes (5,8) this is:

$$d_{58}(10) = 60$$

If node 10 is to be joined to depot 1, its best place in the route is between points 4 and 8 since:

$$d_{48}(10) = \min\left[d_{45}(10); \quad d_{48}(10); \quad d_{58}(10)\right]$$

In the same manner, we examine the increase in route lengths starting from depot 2 and depot 3 if node 10 is added to them. For depot 3 we have:

$$d_{67}(10) = 115 \quad d_{69}(10) = 98 \quad d_{6\,12}(10) = 91$$

We also have for depot 3:

$$91 = d_{6\,12}(10) = \min\left[d_{67}(10); \quad d_{69}(10); \quad d_{6\,12}(10)\right]$$

Depot 2 is only joined to node 11. If node 10 is added to depot 2, the route length would increase by:

$$d_{11}(10) = d_{2,11} + d_{11,10} + d_{10,2} - 2d_{2,11} =$$

$$d_{11,10} + d_{10,2} - d_{2,11} = 47 + 28 - 32 = 43$$

Since:

$$d_{48}(10) = \min\left[d_{48}(10); \quad d_{11}(10); \quad d_{6\,12}(10)\right]$$

we conclude that node 10 should be joined to depot 1 and we add it to the route starting from depot 1 in-between points 4 and 8.

The joining up of nodes to depots is shown in Figure 89.

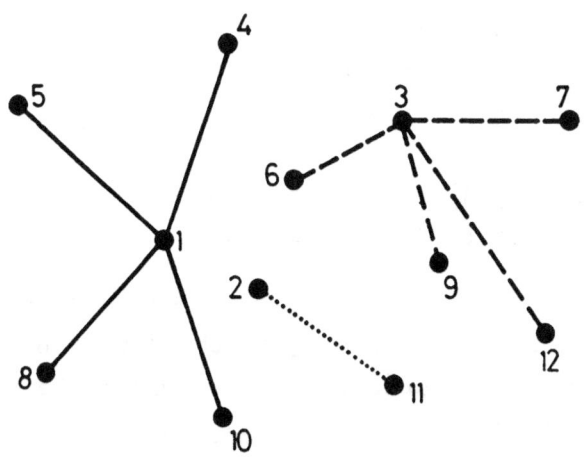

FIGURE 89. Joining nodes to depots.

3.14. **Method to determine the minimum number of vehicles needed to service a given schedule on the transportation network**

One of the basic problems come up against when designing schedules is that of determining the minimum number of vehicles needed to service a given schedule. This problem is relatively easy to solve on smaller transportation networks. However, for larger networks, the optimal solution must be chosen from the very large number of permissable solutions. When solving the problem of determining the minimum number of planes needed to service a given schedule, A. Levin[31] explained a methodological process in 1971 which is also applicable to other forms of public transportation.

Before going into A. Levin's method, we should first discuss certain concepts which are necessary to better understand this process.

An oriented network or oriented graph which does not contain a cycle is called an <u>acyclic oriented graph</u>.

Graph $G(N,A)$ whose set of nodes N can be divided into two subsets S and T so that

$$S \cup T = N \qquad S \cap T = \emptyset$$

and the branches belonging to the set of branches A lead from the nodes of set S towards the nodes of set T is called a <u>bipartite graph</u>.

Bipartite graphs will be denoted as

$$G = \left[S, T; A^* \right]$$

The term decomposing the acyclic oriented graph into chains means dividing the set of nodes N into subsets of nodes which do not have common element and which form chains with the corresponding branches which connect them. We would note that one isolated node can represent a chain as well, which means that acyclic oriented graph $G(N,A)$ can always be decomposed into $|N|$ number of chains, each one made up of only one node ($|N|$ is the number of nodes in set N). Figure 90 shows an acyclic oriented graph whose set of nodes contains nodes $x_1, x_2, \ldots, x_{20}$.

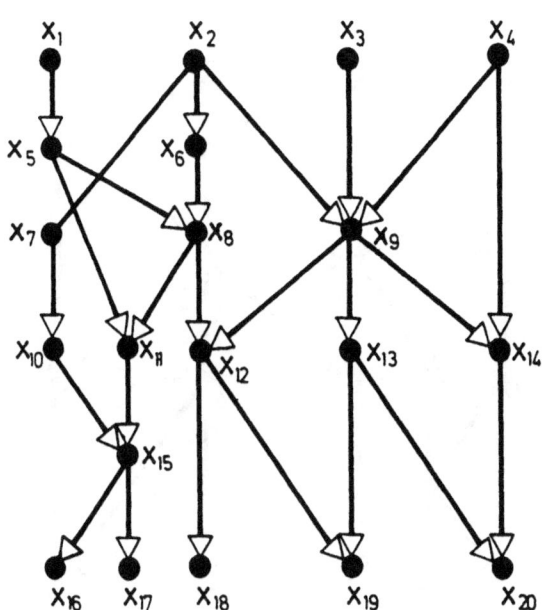

FIGURE 90. An acyclic oriented graph.

The decomposition of this graph is presented in Figure 91. The graph is decomposed into 9 chains. Three chains are made up of only one node (chain (x_6), chain (x_{17}) and chain (x_{19})).

An acyclic oriented graph can be decomposed into chains in several ways. It is clear that the larger the number of nodes included into individual chains, the smaller the number of chains into which the graph is decomposed. The following question appears when decomposing an acyclic oriented graph into chains : What is the smallest number of chains into which an acyclic oriented graph can be decom-

158 TRANSPORTATION NETWORKS

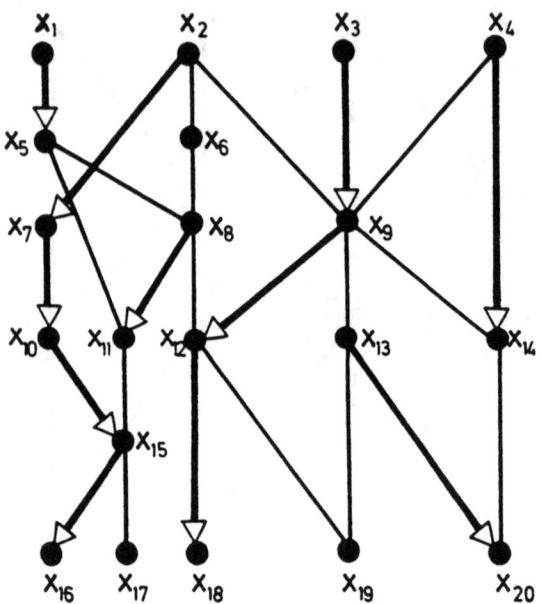

FIGURE 91. Decomposition of an acyclic oriented graph.

posed? However, before answering this question, let us return to our problem and show the connection between determining the minimum number of vehicles needed to service a given schedule on the transportation network and determining the minimum number of chains into which an acyclic oriented graph can be decomposed. We note Figure 92.

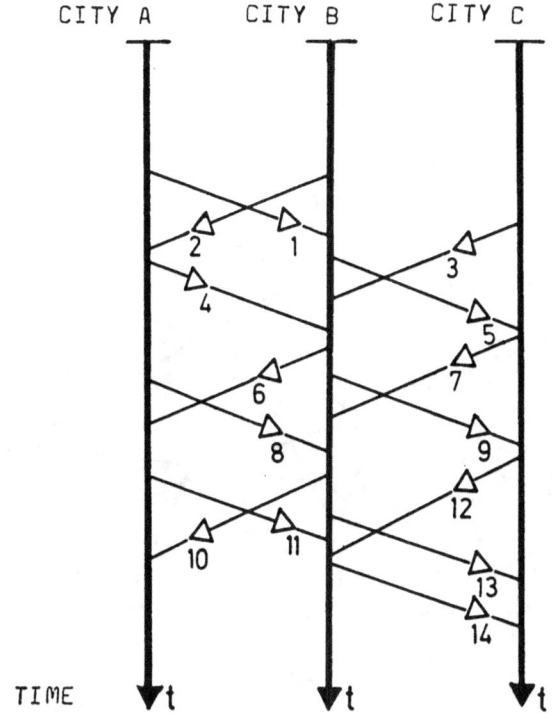

FIGURE 92. Space-time diagram.

Figure 92 shows a space-time diagram with 14 flights (trips) to be carried out between cities A and B, cities B and C, cities C and B and cities B and A. It is obvious that we can distribute the vehicles to carry out the 14 planned flights in different ways. For example, a vehicle can take flight 1, then flight 5, flight 7 and finally flight 10. If we understand flights to be nodes of the network, then chain (1, 5, 7, 10) represents the route taken by the vehicle when carrying out planned flights (trips) 1, 5, 7 and 10. Figure 93 shows a network in which nodes represent planned flights (trips) to be made.

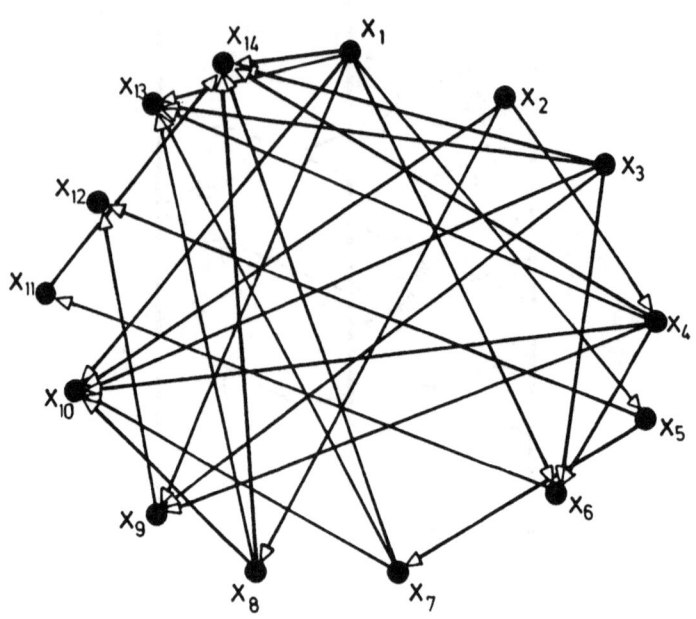

FIGURE 93. Network in which nodes represent planned flights - network $G(N,A)$.

In this network, a branch is directed from node x_i towards node x_j only if flight (trip) x_j can be carried out after flight (trip) x_i. (Flight (trip) x_j can be made after flight (trip) x_i if flight (trip) x_j starts in the city where flight (trip) x_i finishes and if the planned departure time of flight (trip) x_j is after the finishing time of flight (trip) x_i).

VEHICLE ROUTING PROBLEMS ON NETWORKS

Since chains represent vehicle routes, the minimum number of vehicles needed to service a given schedule on the transportation network equals the minimum number of chains into which the acyclic oriented graph can be decomposed, with each node representing flights (trips) to be made.

Therefore, by discovering the answer to our question on the minimum number of chains into which an acyclic oriented graph can be decomposed, we also answer the question on the least number of vehicles needed to service a given schedule.

Let us examine acyclic oriented graph $G(N,A)$. We designate the number of chains into which the graph is decomposed with $|C|$. The chains are denoted respectively by $k = 1, 2, \ldots, |C|$. The number of nodes belonging to the k-th chain is denoted by n_k. The total number of nodes in graph G is denoted by $|N|$. Since every node belongs to only one chain, we have :

$$n_1 + n_2 + \ldots + n_{(c)} = |N|$$

and further :

$$|N| = \sum_{k=1}^{|C|} n_k = \sum_{k=1}^{|C|} n_k + (C-C)$$

$$|N| = \sum_{k=1}^{|C|} (n_k - 1) + |C|$$

The number of branches in every chain is 1 less than the number of nodes in the chain. Therefore, if n_k is the number of nodes in chain k, then $n_k - 1$ is the number of branches in chain k. It is clear that :

$$\sum_{k=1}^{|C|} (n_k - 1)$$

is the number of branches belonging to the chains into which graph G is decomposed.

We denote this number with $|D|$. This means :

$$|D| = \sum_{k=1}^{|C|} (n_k - 1)$$

or :

$$|N| = |D| + |C|$$

Since the number of nodes $|N|$ of graph G is fixed, we can minimize the number of chains $|C|$ into which graph G is decomposed by maximizing the number of branches $|D|$ which belong to the chains into which graph G is decomposed.

We construct bipartite graph $G(S,T;A^*)$ which corresponds to graph $G(N,A)$ shown in Figure 93. The corresponding bipartite graph is shown in Figure 94.

Since, for example, flight (trip) x_5 can be made after flight (trip) x_1, there is a branch in the corresponding bipartite graph $G(S,T;A^*)$ which joins node s_1 with node t_5.

We assume that the capacity of every branch in the bipartite graph $(s_i, t_j) \in A^*$ equals 1. If branch (x_i, x_j) from starting graph G belongs to one of the chains into which graph G has been decomposed, then we note that a flow with a value of 1 goes through the corresponding branch on the bipartite graph. If branch (x_i, x_j) is not part of any of graph G's chains, then we note that there is no flow through the corresponding branch (s_i, t_j) on the bipartite graph, or that the flow value equals 0. It is obvious that the opposite also holds true.

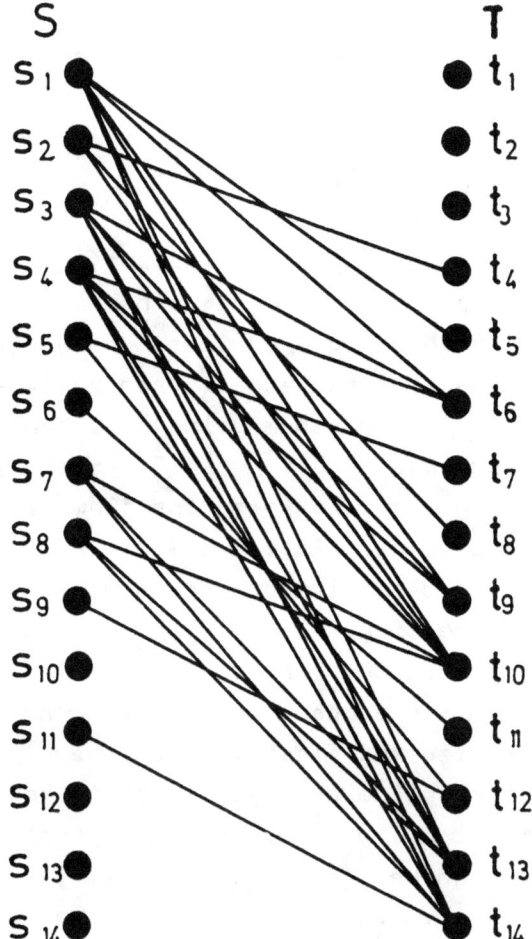

FIGURE 94. Bipartite graph $G(S,T;A^*)$.

If the bipartite graph $G(S,T;A^*)$ contains a flow along branch (s_i, t_j), then branch (x_i, x_j) of graph $G(N,A)$ is part of a chain into which graph G has been decomposed. This means that the total number of branches belonging to graph G's chains, $|D|$, equals the total number of flows going through the bipartite graph. Since some branches which are part of a chain belong to only that chain and some don't belong to any chain, it is clear that a flow with the greatest value of 1 can appear from all source nodes s_1, and that flows with the greatest value of 1 can arrive at all sink nodes t_j.

By minimizing $|C|$ or maximizing $|D|$ we maximize the total flows going through binartite graph $G(S,T;A^*)$, keeping in mind that a flow with a maximum value of 1 can appear from every source s_i and a flow with a maximum value of 1 can arrive at every sink t_j.

E x a m p l e : Maximize the flows in the example used to calculate the smallest amount of vehicles needed to service the schedule shown in the space-time diagram in Figure 92. Figure 95 shows our bipartite graph one more time in which there are only branches starting from node s_1.

We maximize the flows through the bipartite graph in the following way. We start from node s_1 and construct all branches starting from this node. These are branches (s_1, t_5), (s_1, t_6), (s_1, t_9), (s_1, t_{10}), (s_1, t_{13}) and (s_1, t_{14}). We allot a flow value of 1 to the first branch, branch $(s_1, t_5$ Since a flow with the highest value of 1 can appear from node s_1, this means that branches (s_1, t_6), (s_1, t_9), (s_1, t_{10}), (s_1, t_{13}) and (s_1, t_{14}) are left without a flow. The branch allotted the flow is denoted by a heavier line. We also note that in the future all branches arriving at node t_5 will be without a flow since node t_5 has already received a flow of 1.

VEHICLE ROUTING PROBLEMS ON NETWORKS 165

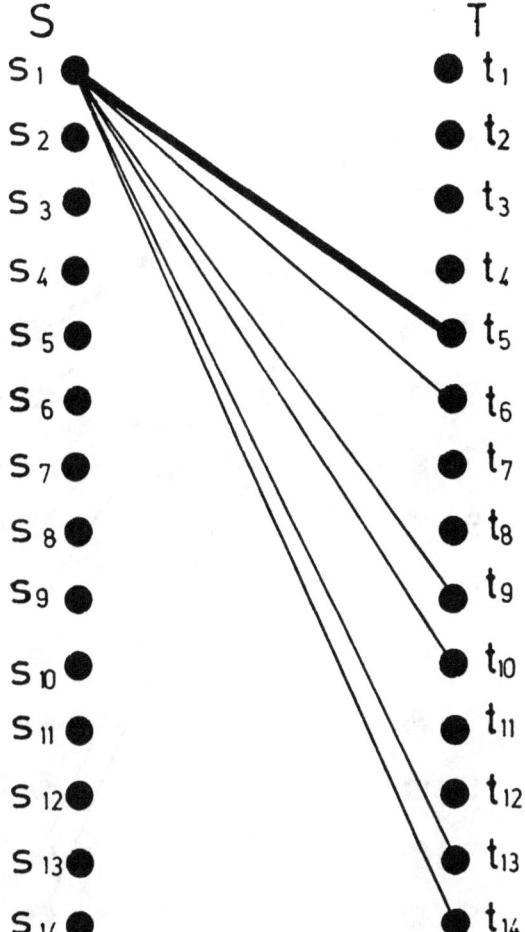

FIGURE 95. Bipartite graph with branches from node s_1.

Now we go to node s_2. We construct branches (s_2, t_4), (s_2, t_8) and (s_2, t_{10}) from this node. The first branch, branch (s_2, t_4) is allotted a flow of 1. Branches (s_2, t_8) and (s_2, t_{10}) will be left without flows as will all future branches

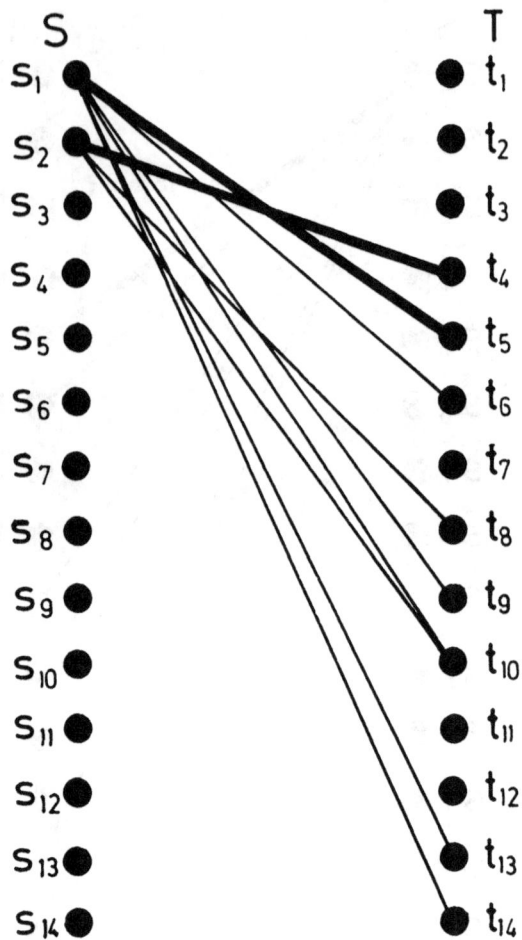

FIGURE 96. Bipartite graph with branches from nodes s_1 and s_2.

which enter node t_4 (Figure 96).

The next node is s_3. Branches starting from this node are branches (s_3, t_6), (s_3, t_9), (s_3, t_{10}), (s_3, t_{13}) and (s_3, t_{14}). Branch (s_3, t_6) is allotted a flow with value 1 and the other branches

VEHICLE ROUTING PROBLEMS ON NETWORKS 167

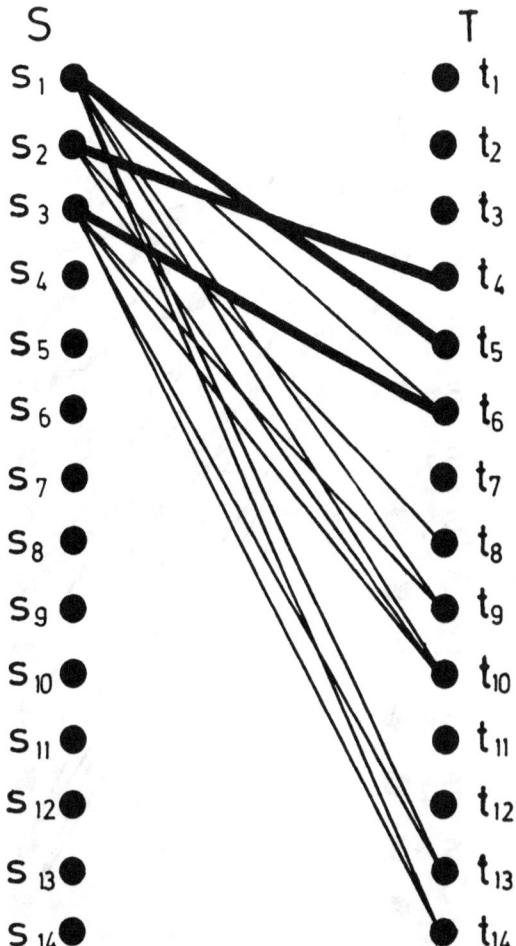

FIGURE 97. Biparte graph with branches from nodes s_1, s_2 and s_3.

are left without flows (Fig.97).

We now come to node s_4. Branches starting here include (s_4, t_6), (s_4, t_9), (s_4, t_{10}), (s_4, t_{13}) and (s_4, t_{14}). Up to now, as a rule, we have always allotted the first branch starting from the node

168 TRANSPORTATION NETWORKS

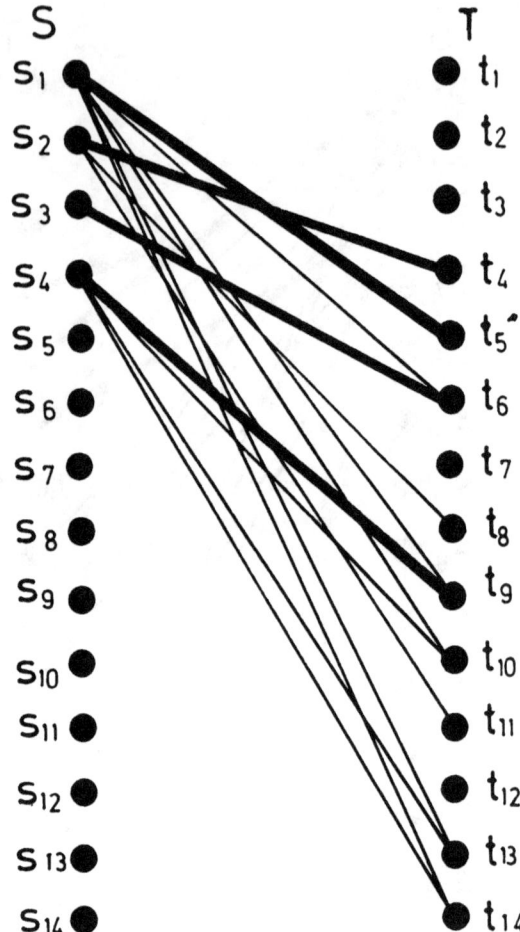

FIGURE 98. Bipartite graph with branches from nodes s_1, s_2, s_4 and s_4.

with a flow value of 1. In this case, branch (s_4, t_6) cannot be allotted a flow since branch (s_3, t_6) arrives at node t_6 with a flow of 1. Therefore, we allot the flow to branch (s_4, t_9). Branches (s_4, t_{10}), (s_4, t_{13}) and (s_4, t_{14}) are left without flows (Fig. 98).

VEHICLE ROUTING PROBLEMS ON NETWORKS 169

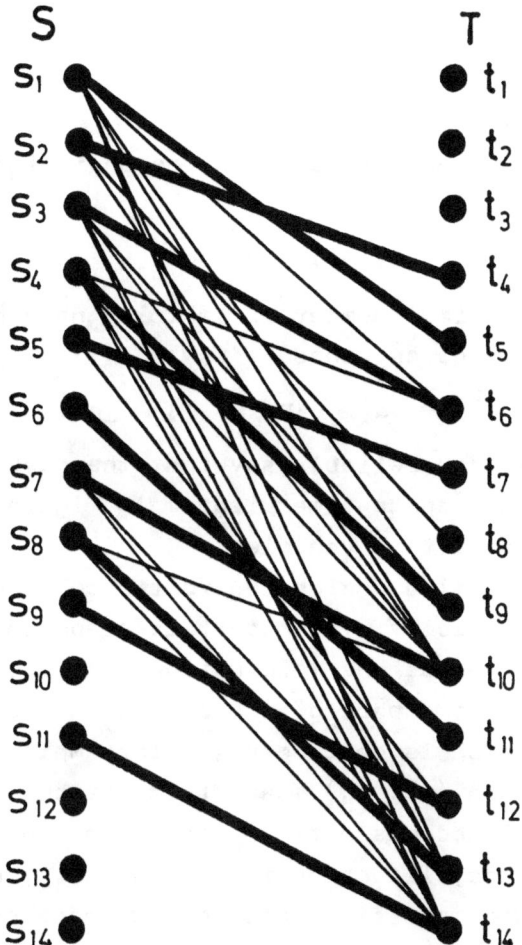

FIGURE 99. Final solution.

Continuing with our procedure, we allot flows of value 1 to branches (s_5, t_7), (s_6, t_{11}), (s_7, t_{10}), (s_8, t_{13}), (s_9, t_{12}) and (s_{11}, t_{14}). The other branches in the bipartite graph will have intensity flows of zero. The final solution is shown on Figure 99.

As we can see from Fig. 99, the bipartite graph

has 10 branches with a flow value of 1. This means that the maximum number of total flows through the bipartite graph is :

$$|D| = 10$$

Since the number of nodes in the beginning graph was :

$$|N| = 14$$

then the **minimum** number of vehicles needed to service the given schedule is :

$$|C| = |N| - |D| = 14 - 10 = 4$$

So, we can reliably say that the given schedule shown on the space-time diagram in Figure 92 can be serviced with 4 vehicles.

We must also design the routes to be serviced by these vehicles. The bipartite graph shown in Fig. 99 gives us the information concerning which branches become part of the chains into which the acyclic graph is decomposed. For example, the flow along branch (s_1, t_5) means that branch (x_1, x_5) in graph G becomes part of one of the chains into which the acyclic graph is decomposed. The acyclic graph's decomposition into chains, made based on Fig. 99 is shown in Figure 100.

Figure 101 shows the routes of the 4 vehicles.

VEHICLE ROUTING PROBLEMS ON NETWORKS 171

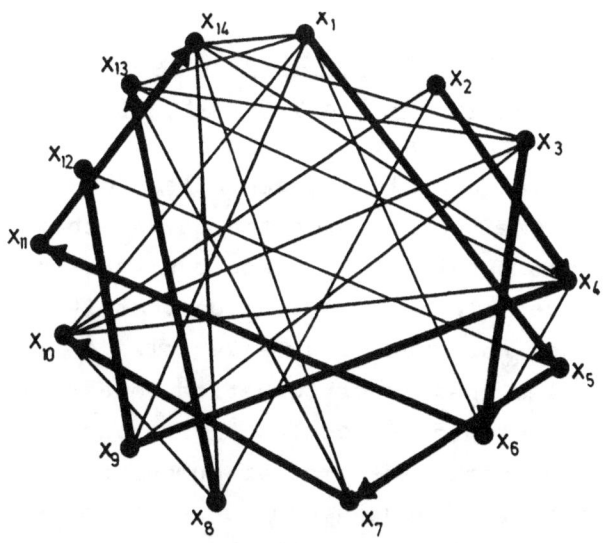

FIGURE 100. Decomposition of acyclic graph into chains.

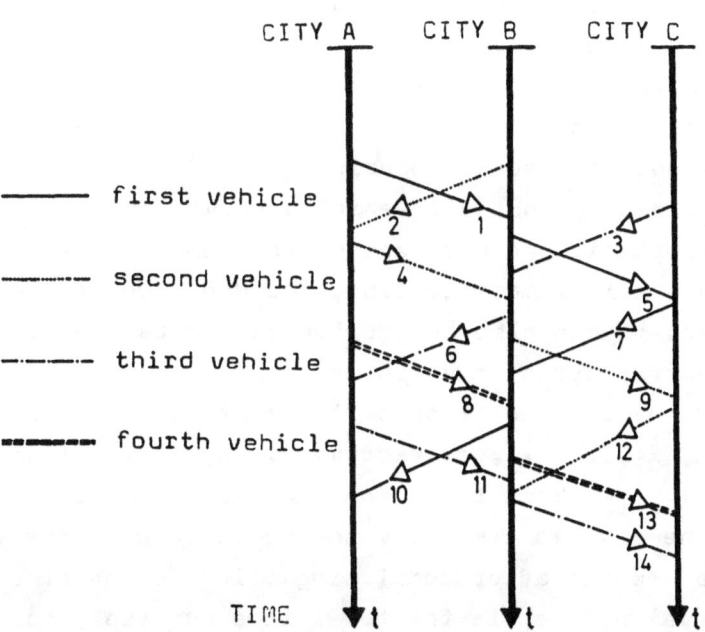

FIGURE 101. Four vehicle routes.

3.15. Optimal dispatching strategy on a transportation network after a schedule perturbation

It often happens (especially in air traffic) that one or several vehicles is cancelled for technical reasons at the beginning of the day so that the carrier must service the schedule with a smaller number of vehicles. It is obvious that if the carrier does not have reserve vehicles to replace the cancelled ones, a perturbation appears in planned dispatching schedules throughout the transportation network. The smaller number of vehicles servicing certain routes causes larger or smaller delays. The problem faced in such a situation can be formulated in the following manner : when one or several vehicles has been cancelled due to technical reasons, design a new schedule to minimize delays throughout the entire network. The method for solving this important problem was given by D. Teodorovic and S. Guberinic in 1983.[55]

We assume that all vehicles servicing the transportation network are of the same capacity. (With a slight modification, the model we are about to explain can also be applied to fleets made up of several types of vehicles).

We denote the flights (trips) to be carried out by $x_1, x_2, \ldots, x_k$, respectively. For every flight (trip) x_i we know the planned departure time a_i and the planned time b_i when the vehicle is ready for a new trip after completing trip x_i. The time interval $b_i - a_i$ is the total time necessary for the vehicle to complete trip x_i and be prepared for a new trip after completing trip x_i. We also know the number of passengers p_i in the vehicle during trip x_i. (This number is most often known in advance in air traffic. If information on the number of passengers p_i on trip x_i cannot be obtained in advance, then the mean number of passengers taking different flights (trips) should be used).

When designing a new schedule, the following condition must be taken into consideration : no vehicle may depart earlier than planned by the old schedule which has now been disturbed.

In other words, if we denote by c_i the departure time of trip x_i by the new schedule, the following must be satisfied :

$$c_i \geq a_i, \qquad i = 1,2,\ldots, k$$

We denote departure time by the new schedule with c_i. We denote by d_i the time by the new schedule when the vehicle is ready for a new trip after completing x_i. It is clear that :

$$d_i = c_i + (b_i - a_i), \qquad i = 1,2,\ldots,k$$

The delay time e_i of flight (trip) x_i is :

$$e_i = c_i - a_i, \qquad i = 1,2,\ldots, k$$

When designing a new schedule, the number of passengers p_i on individual flights (trips) x_i must certainly be taken into consideration. For example, if a plane is 2 hours late, it is not the same if there are 30 passengers or 100 passengers. Total time loss (caused by delay) on flight(trip) x_i is :

$$w_i = e_i p_i \qquad i = 1,2,\ldots,k$$

Total time loss W on the entire transportation network is :

174 TRANSPORTATION NETWORKS

$$W = \sum_{i=1}^{k} w_i$$

Since we wish to minimize total time loss on the entire transportation network, the problem can be mathematically formulated as follows : determine departure times $c_1, c_2, \ldots, c_k$ of flights (trips) $x_1, x_2, \ldots, x_k$ so that :

$$W = \sum_{i=1}^{k} (c_i - a_i) \cdot p_i \longrightarrow \min$$

under the condition that :

$$c_i \geq a_i \qquad i = 1, 2, \ldots, k$$

Let us examine network $G(N,A)$ shown in Figure 102. Flights (trips) x_i to be made represent the nodes of network $G(N,A)$, i.e. :

$$x_i \in N \qquad i = 1, 2, \ldots, k$$

We denote by n the number of vehicles the old schedule was designed for and by m the number of vehicles which are cancelled due to technical reasons. The number of vehicles left is n - m. We denote artificial flights (trips) by $y_1, y_2, \ldots, y_{n-m}$ and assume that each of the remaining vehicles must also make one artificial flight (trip) after which the vehicle is no longer in service. (The reason for introducing artificial flights (trips) will be explained later).

This means that nodes y_j also belong to the set of nodes N of network $G(N,A)$, i.e. :

$$y_j \in N \qquad j = 1, 2, \ldots, n-m$$

Branches (x_i, x_j) going from node x_i towards node x_j belong to the set of branches A of the network $G(N,A)$ if flight (trip) x_i is completed in the city flight (trip) x_j departs from. Let nonoriented branches (x_i, y_j) connecting all nodes x_i with all

VEHICLE ROUTING PROBLEMS ON NETWORKS 175

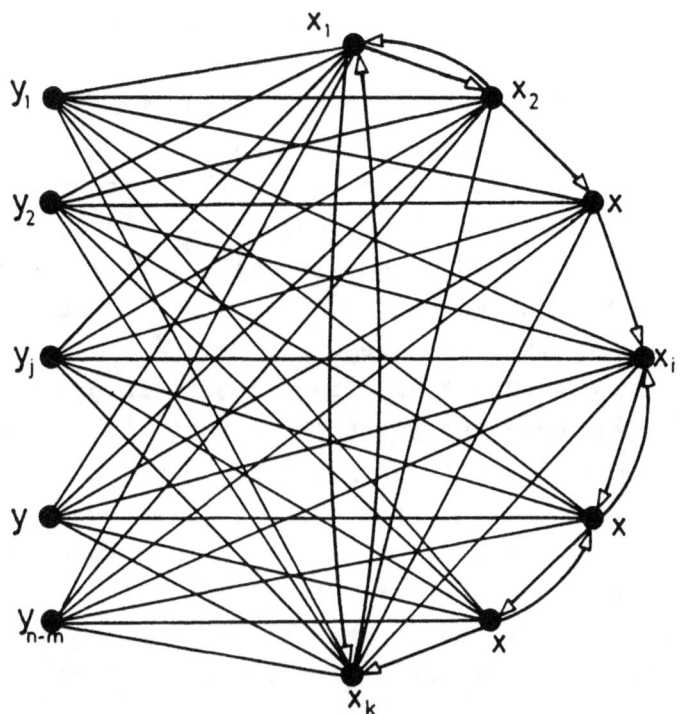

FIGURE 102. Network $G(N,A)$ in which nodes represent flights to be made.

nodes y_j also belong to the set of branches A. Flights (trips) of the remaining $n - m$ vehicles can be represented by a chain q in graph G:

$$q = \left[(x_o,x_b),\ (x_b,x_c),\ \ldots\ (x_r,y_1),\ (y_1,x_h),\ \ldots \\ \ldots (x_p,x_q),\ (x_q,y_j)\right]$$

The first vehicle will carry out fictitious flight (trip) x_o, then flight (trip) x_b, x_c, etc. The next-to-last flight (trip) made by the first vehicle is flight (trip) x_r, and the last is fictitious flight (trip) y_1. This is the last trip made by the first vehicle since after making the fictitious flight (trip), the vehicle is no longer in service. The second vehicle will first make flight x_h, etc. Now we see why we have introduced fictitious nodes into the discussion. These nodes serve as boundaries between groups of flights (trips) made by different vehicles.

It is clear that different chains in graph $G(N,A)$ can represent different flight (trip) combinations made by different vehicles. Each chain always ends in one of the nodes y_j since each vehicle must make one fictitious flight (trip) at the end.

We denote by $d(x_i, x_j)$ the length of branch (x_i, x_j). The length of this branch represents the total time loss of passengers on flight (trip) x_j which is made after flight (trip) x_i.

Since nodes y_j are fictitious, the length of all branches (x_i, y_j) and (y_j, x_i) is zero. Total time loss on the entire transportation network W is actually equal to the length of chain q. This means that:

$$W = \sum_{(x_i,x_j)\in q} d(x_i,x_j) + \sum_{(x_i,y_j)\in q} d(x_i,y_j) + \sum_{(y_i,x_j)\in q} d(y_i,x_j)$$

We want to find at least one (there may be several) of the chains in graph $G(N,A)$ with the shortest total length, while each node x_i and y_j is included in the chain only once.

VEHICLE ROUTING PROBLEMS ON NETWORKS 177

This problem can be solved using the branch-and-bound method which was discussed in detail in the chapter dealing with the traveling salesman problem.

We note length $d(x_i, x_j)$ of branch (x_i, x_j). We already ascertained that this length represents the total time loss of passengers on flight (trip) x_j which is made after flight (trip) x_i. Passengers on flight x_j do not have any time loss if the time flight x_i is completed (time d_i) by the new schedule is before the time that flight x_j is to depart by the old schedule (time a_j).

However, if the time flight x_i finishes by the new schedule (time d_i) is after the time when flight x_j is to depart (time a_j), then the passengers in flight x_j have time loss $(d_i - a_j) \cdot p_j$. This means that :

$$d(x_i, x_j) = \begin{bmatrix} 0, & za\ d_i \leq a_j \\ (d_i - a_j) \cdot p_j & za\ d_i > a_j \end{bmatrix}$$

and that :

$$d(x_i, y_j) = d(y_j, x_i) = 0 \qquad \begin{array}{l} i = 1,2,\ldots,k \\ j = 1,2,\ldots,n-m \end{array}$$

and

$$d(x_o, x_i) = 0 \qquad i = 1,2,\ldots, k$$

To any part of chain q $x_o, \ldots, x_i$ or $x_o, \ldots, y_j$ we add number q_{s,x_i} or q_{s,y_j} which represents the length of all the branches contained in the chain part we are examining. Index s represents the number of nodes contained in the chain part, and index x_i or y_j is the last node on the chain section under consideration.

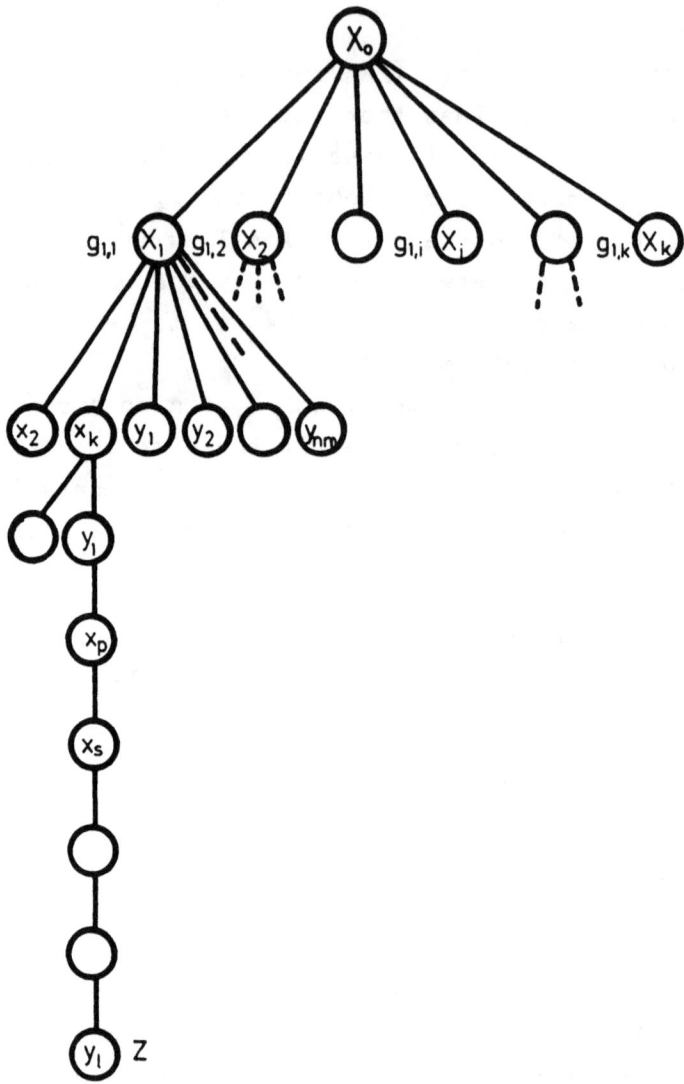

FIGURE 103. Optimal solution obtained using the branch-and-bound method.

It is clear that : $q_{s-1,x_r} = q_{s,y_s} = q_{s+1,x_k}$

$q_{1,x_i} = 0 \qquad i = 1,2,\ldots, k$

As in the problem of the traveling salesman, we will find the optimal solution (the chain with the shortest length) by applying the branch-and-bound method (Figure 103).

We denote the length of the shortest chain by Z, discovered to the moment in which we observe the tree we are branching. Before further branching from any node x_i located on the s-th level, we check whether : $q_{s,x_i} < Z$

If $q_{s,x_i} \geq Z$ we do not branch any further from node x_i on the s-th level since we surely cannot get a better solution than the one we already have.

The branch-and-bound procedure is described in detail in the chapter dealing with the traveling salesman problem. The efficiency of this method when designing a new schedule on a transport network with a perturbation in the old schedule is shown in the following example.

E x a m p l e : The fleet is composed of three aircraft of the same capacity. These 3 aircraft should make 8 flights a day. The space-time diagram on Figure 104 shows the planned flights to be made and also indicates the planned departure times of the flights.

The three airplanes are denoted by a, b and c. The flights to be flown are denoted by x_1, $x_2,\ldots,x_8$. Plane b (shown by the thicker line) which should have flown route $A_1 - A_3 - A_1$ is cancelled for technical reasons. Design a new schedule to be

180 TRANSPORTATION NETWORKS

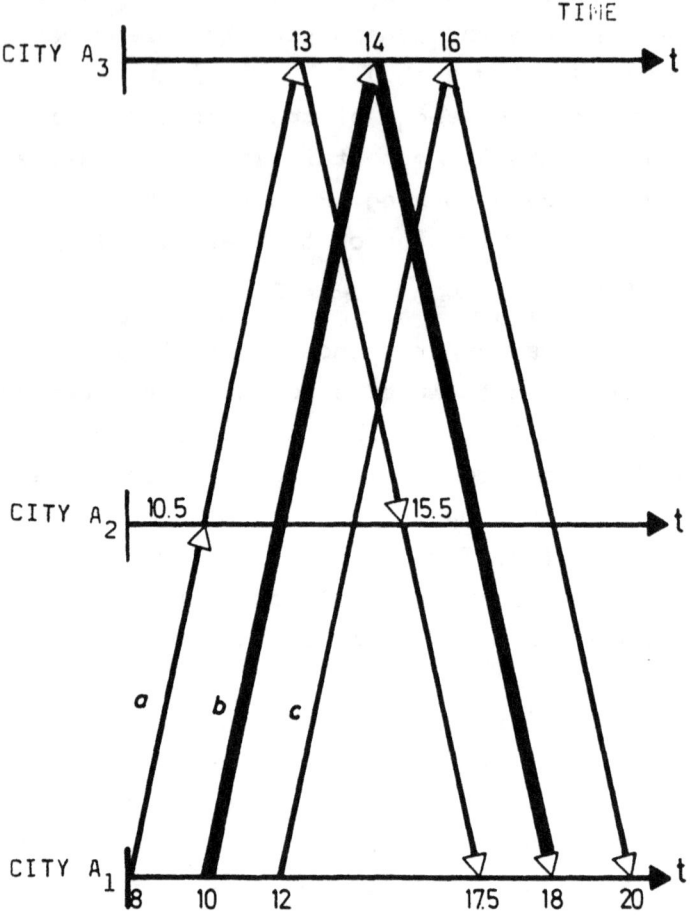

FIGURE 104. Space-time diagram.

serviced by planes a and c so that delays on the entire route network are minimized. Planned departure times by the old schedule and the number of passengers per flight are given in Table XIV.

TABLE XIV. Planned departure times and number of passengers.

Flight x_i	Planned departure time a_i	Planned time when plane ready for new flight after flight x_i b_i	Number of passengers on flight x_i p_i
x_1	8	10	105
x_2	10,5	13	25
x_3	13	13,5	35
x_4	15,5	17,5	110
x_5	10	14	65
x_6	14	18	70
x_7	12	16	25
x_8	16	20	30

Since 2 planes will handle all the traffic, graph $G(N,A)$ will contain 2 fictitious nodes y_1 and y_2. Nodes $x_1, x_2, \ldots, x_8$ belong to set N and they represent flights to be made. The space-time diagram gives us information on the branches which will be in graph $G(N,A)$. For example, branch (x_1, x_2) going from node x_1 towards x_2 will be in graph $G(N,A)$ since flight x_1 is completed in city A_2 and flight x_2 starts from the same city. In the same vein, branch (x_6, x_1) going from node x_6 to node x_1 will also be in graph $G(N,A)$ since flight x_6 finishes in city A_1 and flight x_1 starts from the same city, etc. Graph $G(N,A)$ is shown in Fig. 105.

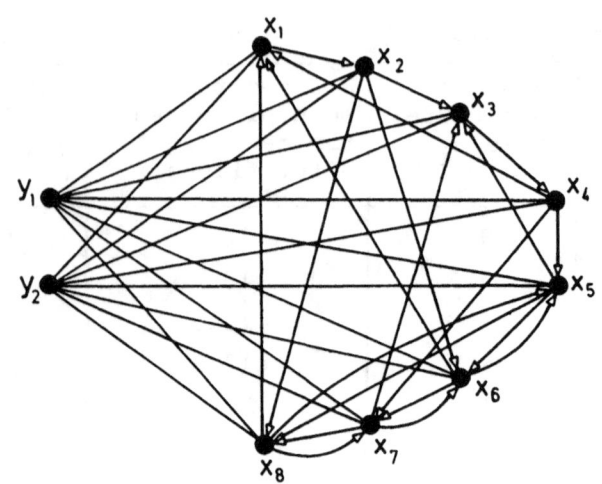

FIGURE 105. Graph G(N,A).

The optimal solution is obtained using the branch-and-bound method. Branching and bounding are done using a HP-9825-A computer. The corresponding computer program is written in HPL language.

The chain with the shortest length is :

$$\left[(x_1,x_2),(x_2,x_3),(x_3,x_4),(x_4,x_7),(x_7,x_8),(x_8,y_1),\right.$$
$$\left.(y_1,x_5),(x_5,x_6)(x_6,y_2)\right]$$

The length of this chain is 302.5 hours.

Fictitious node y_1 limited the flights to be made by planes a and c. Plane a should make flights x_1, x_2, x_3, x_4, x_7 and x_8 (in that order) and plane c should make flights x_5 and x_6.

Table XV gives planned and realized departure times and time losses on the 8 flights.

TABLE XV. Planned and realized departure times and time losses.

Flight x_i	Planned departure time a_i	Realized departure time c_i	Number of passengers on flight x_i p_i	Time loss on flight x_i $W_i \equiv (c_i - a_i)p_i$
x_1	8	8	105	0
x_2	10,5	10,5	25	0
x_3	13	13	35	0
x_4	15,5	15,5	110	0
x_5	10	10	65	0
x_6	14	14	70	0
x_7	12	17,5	25	137,5
x_8	16	21,5	30	165

Figure 106 shows chain sections belonging to plane a and plane c. In other words, these are the flights made by a and those made by c.

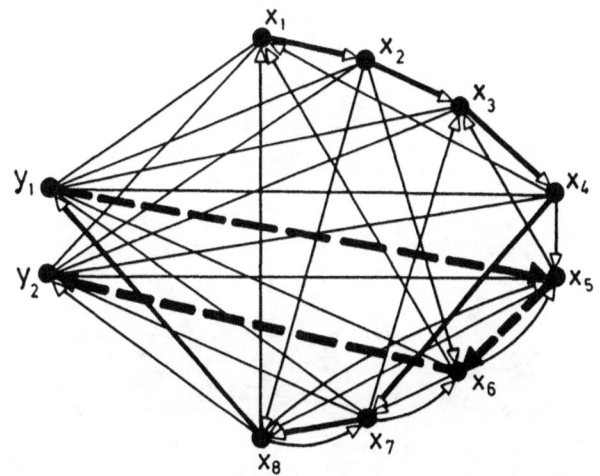

FIGURE 106. Optimal solution.

4 Determining Vehicle Depot Locations

The quality of transportation services and the transportation system's total costs primarily depend on the location of vehicle depots. The position of the depot, i.e. specific facilities on the network where some type of service takes place, also depends on the type of service. For example, it is clear that airports must be located as far as possible from the city center due to ecological reasons. On the other hand, they should not be too far away since this represents a significant decrease in the quality of air transportation service. The locations of public city transportation stations are also the result of an attempt to minimize users' walking distances, while fire fighting brigades, ambulance services and police stations are conditioned by the requirement to minimize the distance to the farthest user. As can be seen, the number and location of specific service facilities on the transportation network are a function of certain criteria and requirements set before the transportation system.

Numerous papers have been published during the past two decades dealing with the problem of locating facilities on the transportation network.

We mentioned above that locating fire brigades, ambulance and police stations are most often stipulated by the requirement to minimize the distance to the farthest user. This type of problem is usually called a center problem and sometimes it is called a minimax problem.

Another group of papers deals with the problem of locating several facilities in order to minimize the average "distance" between facilities and service users. We would note that "distance" can also stand for travel time, travel costs or some other quantity. This type of problem is known an a median problem and it is encountered when designing networks for distribution centers, locations for schools, post offices, shops, etc.

When discussing center and median problems, the assumption is most often made that service demands appear only on the transportation network nodes with each node attributed a certain service demand intensity. It is clear that in a certain number of cases service demands also appear along the transportation network branches. In such cases a number of new nodes are introduced into the discussion which enable the condition to be satisfied that demands for service can only appear at nodes.

4.1. Determining transportation network centers

Since center problems strive to minimize the distance to the farthest user, shortest paths d_{ij} between all pairs of nodes (i,j) must be calculated before applying a suitable algorithm for determining the center of a given transportation network.

Before introducing the algorithm for determin-

DETERMINING VEHICLE DEPOT LOCATIONS

ing the center of a transportation network, we will explain the concepts of local center, node center and absolute center.

Let us now study nonoriented network $G(N,A)$. Let $x \in G$ be some point on the network. We denote the shortest distance between point x and node i by $d_{x,i}$. We denote the distance between point x and the node of network G which is the farthest from point x as :

$$f(x) = \max_{i \in N} d_{x,i}$$

Now let us study some branch $a \in A$. Point x_a on branch a is called the local center of branch a if the following is true for every point $x \in a$:

$$f(x_a) \leq f(x)$$

Node $j^{(c)}$ is called the node center of network G if the following is true for every node $i \in N$:

$$f(j^{(c)}) \leq f(i)$$

Point x_o is called the absolute center of network G if the following is true for every node $x \in G$:

$$f(x_o) \leq f(x)$$

The procedure for determining the absolute center of a network follows from the definitions of local and absolute network center. To this effect, in the first step, local center x_a must be found for each branch a of network G. In the second step, the local center with the smallest value $f(x_a)$ should be chosen, and this is simultaneously the absolute center x_o of network G.

188 TRANSPORTATION NETWORKS

Example : Let us discuss the transportation network in Figure 107.

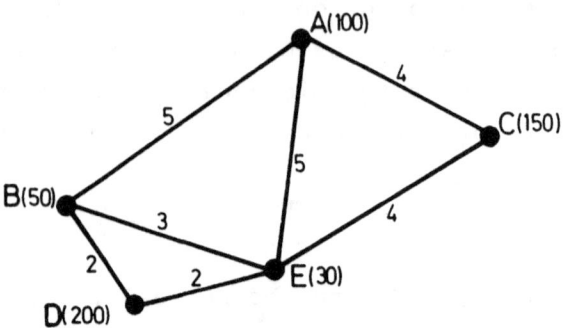

FIGURE 107. Transportation network for which center has to be determined.

Let nodes A, B, C, D and E represent forest-covered islands. The area covered by forests is given in parentheses next to each node. The distance between each node is also given on the figure. A joint airborne fire-fighting brigade is to be designed for these five islands and the optimal location of the fire-fighting brigade must be determined. The optimal location must minimize the greatest distance between potential fire locations and the airborne fire-fighting brigade station.

The fire station can be on only one of the five islands. First, we find the shortest paths between all pairs of nodes and show the distance as elements in the matrix of shortest paths d_{ij}. The node which has the minimum value of the maximum elements of its row is the optimal location for the center. In

DETERMINING VEHICLE DEPOT LOCATIONS 189

this case node E is the optimal location for the center.

$$[d_{ij}] = \begin{array}{c} \\ A \\ B \\ C \\ D \\ E \end{array} \begin{bmatrix} A & B & C & D & E \\ 0 & 5 & 4 & \boxed{7} & 5 \\ 5 & 0 & \boxed{7} & 2 & 3 \\ 4 & \boxed{7} & 0 & 6 & 4 \\ \boxed{7} & 2 & 6 & 0 & 2 \\ \boxed{5} & 3 & 4 & 2 & 0 \end{bmatrix}$$

Now let us assume that nodes A, B, C, D and E represent cities. The number in parentheses next to each node are the number of inhabitants in each city. In this case, the fire-fighting brigade can be located in any spot along the branches of the transportation network.

The procedure for finding the local center in which to place the joint fire station for all five cities is shown in Figures 108 and 109.

Let us study branch (A,B) which has a length of 5 units. For all points x located on branch (A,B) we draw function $d_{x,i}$ for i = A, B, C, D, E. For example, if we put x = 0 in A and x = 5 in B, then:

$$d_{x,B} = 5 - x \qquad \text{for } 0 \leqslant x \leqslant 5$$

$$d_{x,D} = 7 - x \qquad \text{for } 0 \leqslant x \leqslant 5$$

Functions $d_{x,A}$, $d_{x,B}$, $d_{x,C}$, $d_{x,D}$ and $d_{x,E}$ referring to branch (A,B) are shown on Fig. 108.

By definition:

$$f(x) = \max \left[d_{x,A}; \ d_{x,B}; \ d_{x,C}; \ d_{x,D}; \ d_{x,E} \right]$$

H

190 TRANSPORTATION NETWORKS

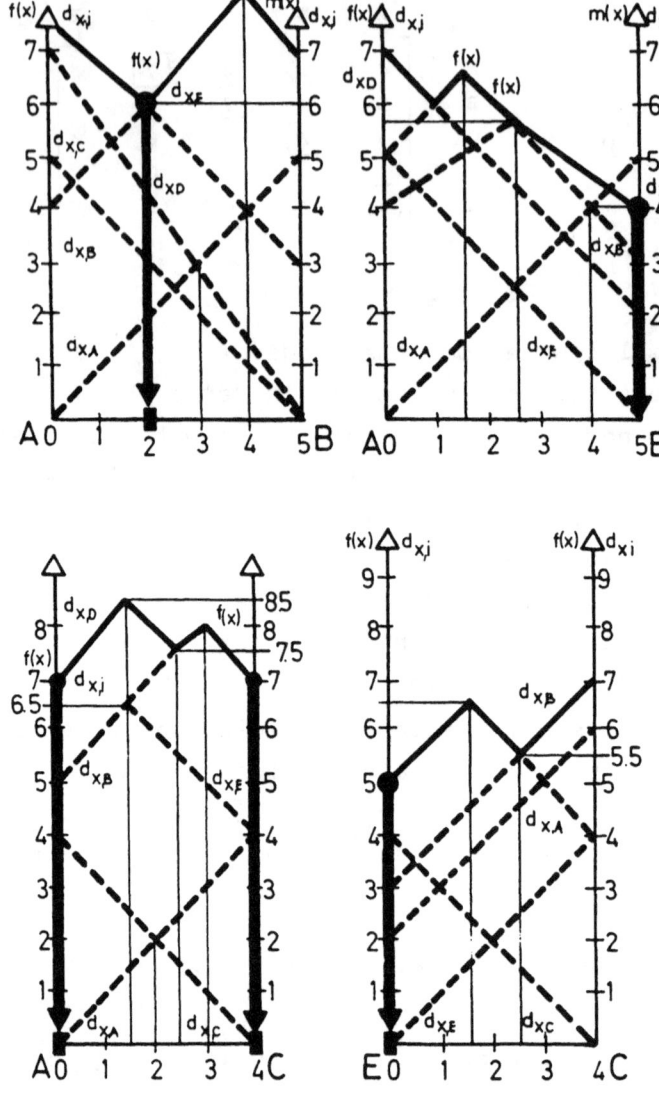

FIGURE 108. Finding local centers.

DETERMINING VEHICLE DEPOT LOCATIONS

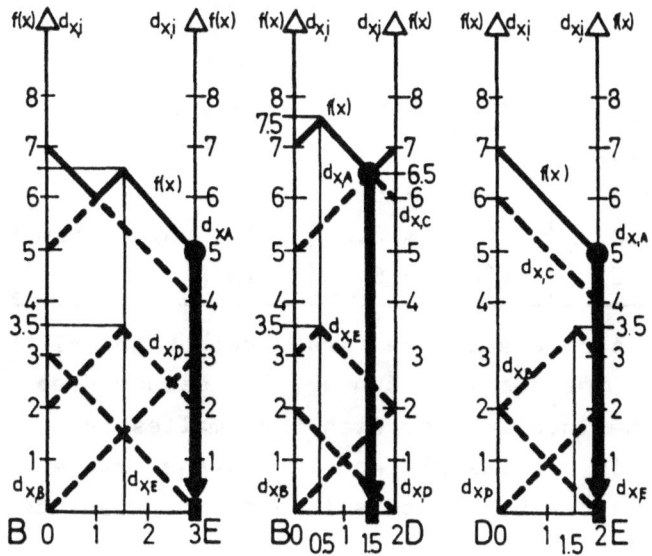

FIGURE 109. Finding local centers.

Function f(x) is drawn on Fig. 108 with a thick line. It is obvious from Fig. 108 that the local center of branch (A,B) is the point which is 2 units away from A and 3 units away from B. For branch (A,B) we have :

$$f(x_a) = 6$$

By applying this procedure to the other branches the following local centers are obtained :

TABLE XVI. Local centers

Branch	Function $f(x)$	Local Center
(A,B)	$f(x_a) = 6$	2 units from A
(A,E)	$f(x_a) = 4$	in E
(A,C)	$f(x_a) = 7$	in A and C
(E,C)	$f(x_a) = 5$	in E
(B,E)	$f(x_a) = 5$	in E
(B,D)	$f(x_a) = 6.5$	1.5 units from B
(D,E)	$f(x_a) = 5$	in E

This completes the first step of the algorithm to find the absolute center. The second step is to find the local center with the smallest value of $f(x_a)$. In our example node E has the smallest value, i.e. $f(x_a) = 5$, and this is also the absolute center.

4.2. The problem of several centers

The problem of several centers was dealt with by Hakimi and other in 1978.[27] This problem of determining the location of several centers is comprised of the following :

Let us study nonoriented network $G(N,A)$. Let $X_p = (x_1, x_2, \ldots, x_p)$ be the set of p points in network G. We denote by $d(i,X_p)$ the shortest distance between node i and any point $x_1, x_2, \ldots, x_p$:

$$d(i,X_p) = \min d_{i,x_j} \qquad x_j \in X_p$$

We denote by $f(X_p)$ the largest of all the shortest distances between nodes of network G and any point $x_1, x_2, \ldots, x_p$:

$$f(X_p) = \max d(i,X_p) \qquad i \in N$$

The set of p points X_p^* is called the p-center of network G if for every $X_p \in G$, the following is true:

$$f(X_p^*) \leq f(X_p)$$

It should be noted that thus far in our discussion of transportation network centers we have not considered the intensity of service demand u_i in the nodes of network $i \in N$. Quantity u_i can also stand for the weight or significance attached to node i. When using quantity u_i in center problems, the following expression should be used instead of the shortest distance $d(i, X_p)$ between node i and any point $x_1, x_2, \ldots, x_p$:

$$u_i \, d(i, X_p)$$

Analogous to this, $F(X_p)$ is defined as:

$$F(X_p) = \max \left[u_i \, d(i, X_p) \right] \quad i \in N$$

In this case the set of p points X_p^* is called the p-center of network G, if for every $X_p \in G$ the following is true:

$$F(X_p^*) \leq F(X_p)$$

4.3 The median problem

Median problems are particularly important for transportation activity since they appear when designing different distribution systems. In this case, we try to minimize the average "distance" between facilities where some service is provided and the user of that service.

Let us consider nonoriented network $G(N, A)$ which has n nodes. We denote by a_i the number of service demands from node i. The corresponding service demand intensity u_i from node i is:

194 TRANSPORTATION NETWORKS

$$u_i = \frac{a_i}{\sum_{i=1}^{n} a_i} \quad , \quad \forall_i$$

It is clear that :

$$\sum_{i=1}^{n} u_i = 1$$

It is logical to assume that in most distribution systems each node is serviced from the depot which is closest to it. Let there be p depots in the network, i.e. p facilities where some service is carried out. The set of all facilities is denoted by X_p. We denote by $d(i,X_p)$ the shortest distance between node i and any point $x_1, x_2,\ldots, x_p$ which contains facilities.

The average distance between facilities and users in this network is :

$$L(X_p) = \sum_{i=1}^{n} u_i \, d(i,X_p)$$

It is clear that we are trying to locate facilities in the network so as to minimize the average distance $L(X_p)$ between facilities and users.

The set of points X_p^* in network G is the set of p-medians of network G if the following is true for every $X_p \in G$:

$$L(X_p^*) \leq L(X_p)$$

Hakimi[26a] showed that there is at least one set of p-medians in the nodes of network G, which means that p optimal locations for facilities in the network must be found exclusively in the network nodes. This fact significantly facilitates the procedure for finding p medians since only locations found in nodes must be examined.

4.4. Algorithm to determine one network median

The algorithm for determining one median for non-oriented network $G(N,A)$ is composed of the following algorithmic steps :

Step 1 : Calculate the shortest distances d_{ij} between pairs of nodes (i,j) in network G and show them in a matrix of shortest distances.

Step 2 : Multiply the j-th column of the matrix by the number of service demands a_j from node j. Element $a_j \cdot d_{ij}$ of matrix $(a_j d_{ij})$ is the "distance" covered by users from node j who are serviced in node i.

Step 3 : Divide elements of matrix $(a_j d_{ij})$ by $\sum_{j=1}^{n} a_j$. A new matrix is obtained whose elements are $u_j d_{ij}$. Add up the elements of every row i of matrix $(u_j d_{ij})$. Expression $\sum_{j=1}^{n} u_j d_{ij}$ is the average "distance" covered by users when the facility is in node i.

Step 4 : The node whose row corresponds to the least average "distance" covered by users is the median location.

E x a m p l e : Let us discuss the transportation network shown in Figure 110.

Transportation network nodes are denoted by A, B, C,..., H. Daily service demands are given in parentheses next to each node. All branch lengths are also marked.

196 TRANSPORTATION NETWORKS

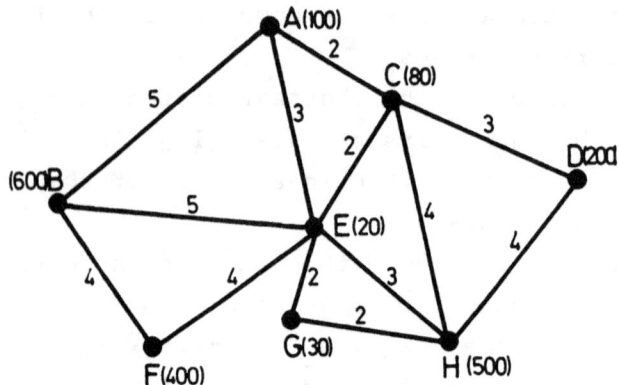

FIGURE 110. Transportation network in which one median is to be found.

The problem is as follows : Where should we locate a service facility so that the average distance covered by service users (located in the nodes) to the facility is minimized?

Based on Hakimi's theory, we can conclude that there are 8 candidate spots in which to locate the facility. These are nodes A, B, C,..., H. Using the shortest path algorithm we calculate shortest paths d_{ij} between all pairs of nodes (i,j) in the transportation network. These shortest distances are given in the following matrix :

DETERMINING VEHICLE DEPOT LOCATIONS 197

$$d_{ij} = \begin{array}{c} \\ A \\ B \\ C \\ D \\ E \\ F \\ G \\ H \end{array} \begin{array}{cccccccc} A & B & C & D & E & F & G & H \\ \left[\begin{array}{cccccccc} 0 & 5 & 2 & 5 & 3 & 7 & 5 & 6 \\ 5 & 0 & 7 & 10 & 5 & 4 & 7 & 8 \\ 2 & 7 & 0 & 3 & 2 & 6 & 4 & 4 \\ 5 & 10 & 3 & 0 & 5 & 9 & 6 & 4 \\ 3 & 5 & 2 & 5 & 0 & 4 & 2 & 3 \\ 7 & 4 & 6 & 9 & 4 & 0 & 6 & 7 \\ 5 & 7 & 4 & 6 & 2 & 6 & 0 & 2 \\ 6 & 8 & 4 & 4 & 3 & 7 & 2 & 0 \end{array}\right] \end{array}$$

In the next step we calculate expression $a_j \cdot d(i,j)$ by multiplying every column of the shortest paths matrix by the number of service demands in node j. Matrix $\left[a_j \cdot d_{ij}\right]$ is as follows:

$$\left[a_j d_{ij}\right] = \begin{array}{c} \\ A \\ B \\ C \\ D \\ E \\ F \\ G \\ H \end{array} \begin{array}{cccccccc} A & B & C & D & E & F & G & H \\ \left[\begin{array}{cccccccc} 0 & 3000 & 160 & 1000 & 60 & 2800 & 150 & 3000 \\ 500 & 0 & 560 & 2000 & 100 & 1600 & 210 & 4000 \\ 200 & 4200 & 0 & 600 & 40 & 2400 & 120 & 2000 \\ 500 & 6000 & 240 & 0 & 100 & 3600 & 180 & 2000 \\ 300 & 3000 & 160 & 1000 & 0 & 1600 & 60 & 1500 \\ 700 & 2400 & 480 & 1800 & 80 & 0 & 180 & 3500 \\ 500 & 4200 & 320 & 1200 & 40 & 2400 & 0 & 1000 \\ 600 & 4800 & 320 & 800 & 60 & 2800 & 60 & 0 \end{array}\right] \end{array}$$

By summing up the rows of matrix $a_j \cdot d(i,j)$ we get the number of kilometers traveled by users if the facility is located in the node whose row is being summed up.

It is clear that the facility should be located in the node whose summed up row is the smallest. By dividing the total sum by the total number of users in the entire transportation network (in this case 1930), we get the average distance covered by one user in order to satisfy his demand in the facility. The following table gives the total number of kilometers traveled and the average distance covered by one user when the facility is located in a specific node in the network :

TABLE XVII. Total number of kilometers traveled

Facility located in node	No. of kilometers traveled	Average distance
A	10170	5.2
B	8970	4.6
C	9760	5.05
D	12620	6.5
E	7620	3.9
F	9140	4.7
G	9660	5
H	9440	4.8

It is clear that the facility should be located in node E since the average distance of 3.9 kilometers is smaller than all other average distances.

4.5. Algorithm for finding k medians

Let us now look at the transportation network shown on Figure 110. Determine the location of two medians for this network.

It is clear (from Hakimi's theory) that the medians must be located in 2 nodes from the set of nodes (A, B, C, D, E, F, G, H). There is a total of 28 possibilities for locating the two medians. We assume that the user will always use the services

DETERMINING VEHICLE DEPOT LOCATIONS 199

of the nearest facility. So, for example, the number of kilometers traveled by users from node D when the facility is located in nodes A and C is :

$$\min \left[a_D \cdot d_{A,D}; \; a_D \cdot d_{C,D} \right] = \min \left[1000; \; 600 \right] = 600$$

Based on matrix $(a_j \cdot d_{ij})$ the total number of kilometers traveled by users can be calculated for any of the 28 possibilities.

The total number of kilometers traveled for each of the 28 possible depot locations is given in the following table :

TABLE XVIII Total number of kilometers traveled

No.	Node pair with depots	No. of km.	No.	Node pair with depots	No. of km
1.	A,B	5970	15	C,E	6960
2.	A,C	8160	16	C,F	5560
3.	A,D	8170	17	C,G	8640
4.	A,E	7380	18	C,H	7500
5.	A,F	6770	19	D,E	6620
6.	A,G	7600	20	D,F	5400
7.	A,H	6880	21	D,G	8380
8.	B,C	4560	22	D,H	8460
9.	B,D	4620	23	E,F	5420
10.	B,E	4620	24	E,G	7100
11.	B,F	6530	25	E,H	5920
12.	B,G	4660	26	F,G	5460
13.	B,H	3340	27	F,H	4240
14.	C,D	8960	28	G,H	8260

We can see that the least number of kilometers is traveled when the depots are located in nodes B and H. Figure 111 shows the transportation network with its depots in nodes B and H. The nodes

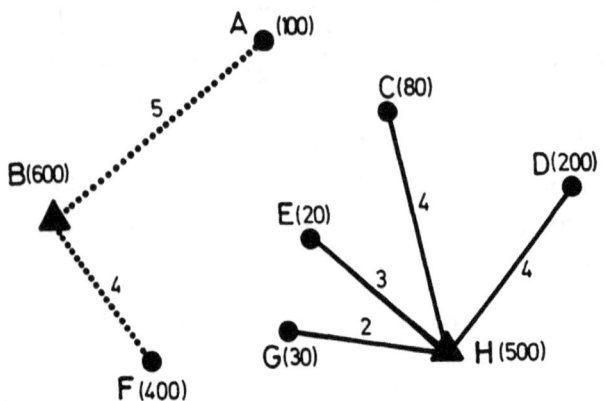

FIGURE 111. Depots B and H.

serviced by each depot are also denoted.

In this case we determined the two medians rather easily. However, when a larger number of medians is to be determined in a transportation network with more nodes, considerable calculating difficulties can arise when this procedure is used. A special efficient algorithm has been developed to determine a larger number of medians.

This efficient algorithm for solving the problem of k medians is given in Larson and Odoni's book.[29]

The algorithm starts with finding the location of one median and a new median is found with each additional step. The algorithm is finished when all k medians have been found. Due to Hakimi's theory, the location of all k medians can be only

DETERMINING VEHICLE DEPOT LOCATIONS 201

on the nodes of the network under observation. We denote by S the set of nodes where medians are (temporarily) located and by m the number of nodes in set S. While searching for the locations of the k medians, m grows from 1 to k. This heuristic algorithm is composed of the following steps :

Step 1 : Let m = 1. Find the location for one median using the appropriate algorithm. Let this median be located in node i. This means that $S = (i)$.

Step 2 : The next median is located in a node from the set N - S so that it achieves the greatest improvement to the objective function. Now let m = m + 1.

Step 3 : We now try to improve the value of the objective function by systematically replacing one of the nodes in set S by one of the nodes in set N - S every time an improvement takes place and when no more improvements can be made to the objective function, go to step 4.

Step 4 : If m = k the algorithm is finished. If not, go to Step 2.

E x a m p l e : Use the described algorithm to find the 2 medians in the transportation network shonw on Fig. 111.

In the first step we determine that the optimal location for one median is node E. For this reason $S = (E)$. In the second step we must compare the values of the objective function for the case when the medians are located in the following nodes :

      A,E,   B,E,   C,E,   D,E,   E,F,   E,G,   E,H.

202 TRANSPORTATION NETWORKS

Based on the values of the elements in matrix $(u_j \cdot d(i,j))$ we get objective function values of 7380, 4620, 6960, 6620, 5420, 7100 and 5920. This means that now $S = (B,E)$. We now compare this solution with solutions :

    A,B,   B,C,   B,D,   B,F,   B,G,   B,H.

Since the value of the objective function is 3340 for the solution $S = (B,H)$ this is now the new solution. The objective function for this solution is now compared to the values for the following solutions :

    A,H,   C,H,   D,H,   E,H,   F,H,   G,H.

After these comparisons, there are no improvements in the objective function. The solution B,H is compared to the solutions for : A,B, B,C, B,D, B,E, B,F and B,G. No improvements can be made to the objective function after these comparisons, either. Since m = 2 = k, we have completed the algorithm with the solution $S = (B,H)$.

This algorithm can be easily made into a computer program which works very quickly and gives a solution close to optimal.

The problem of one or more medians was observed for nonoriented networks. For oriented networks, care must be taken as to which is more desirable, to minimize the distance from the user to the facility, from the facility to the user or whether to minimize the total distance from user to facility and back again.

E x a m p l e : Determine the least number and locations of facilities to provide service for the transportation network shown in Figure 112 so that

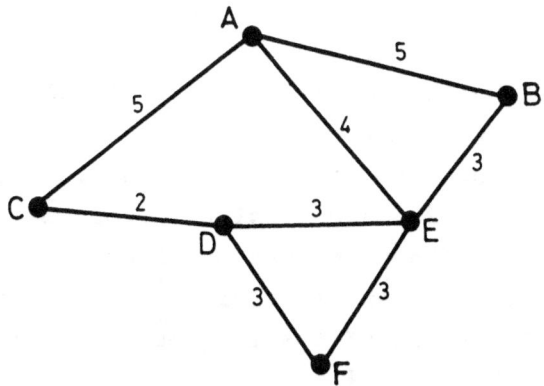

FIGURE 112. Transportation network G(N,A)

the greatest distance between any node and its nearest facility is not more than 4. Facilities can only be located in network nodes.

The least number of facilities needed is denoted by p. At the beginning, p = 1. We first solve the problem of one center for our network. The shortest path matrix between all nodes is :

$$[d_{ij}] = \begin{array}{c} \\ A \\ B \\ C \\ D \\ E \\ F \end{array} \begin{bmatrix} A & B & C & D & E & F \\ 0 & 5 & 5 & 7 & 4 & 7 \\ 5 & 0 & 8 & 6 & 3 & 6 \\ 5 & 8 & 0 & 2 & 5 & 5 \\ 7 & 6 & 2 & 0 & 3 & 3 \\ 4 & 3 & 5 & 3 & 0 & 3 \\ 7 & 6 & 5 & 3 & 3 & 0 \end{bmatrix} \begin{array}{c} \text{Maximum value in row} \\ 7 \\ 8 \\ 8 \\ 7 \\ \text{⑤} \\ 7 \end{array}$$

If only one facility existed to service all nodes, it should be located in node E. The farthest node from node E would then be at a distance of 5 units. However, the problem requires for the greatest distance between any node and its nearest facility to be not more than 4. It is clear that if only one facility existed in the network, this condition could not be satisfied. We increase p by 1. Now, p = 2. We now solve the problem for 2 node centers. The shortest distance between all nodes and their nearest service facility for all possible locations of the two facilities is given in Table XIX.

The problem's condition that the greatest distance between any node and its nearest facility not exceed 4 is fulfilled if the facilities are located in either nodes C and E or in nodes D and E. This means that two facilities are sufficient to provide service to all users from all nodes and also fulfill the pre-set condition. Problems of this type which contain some pre-set requirement are most often called requirement problems.

TABLE XIX. Minimum distances between nodes and the nearest service facility.

Node pairs with facility	Node						Maximum value in row
	A	B	C	D	E	F	
A,B	0	0	5	6	3	6	6
A,C	0	5	0	2	4	5	5
A,D	0	5	2	0	3	3	5
A,E	0	3	5	3	0	3	5
A,F	0	5	5	3	3	0	5
B,C	5	0	0	2	3	5	5
B,D	5	0	2	0	3	3	5
B,E	4	0	5	3	0	3	5
B,F	5	0	5	3	3	0	5
C,D	5	6	0	0	3	3	6
C,E	4	3	0	2	0	3	④
C,F	5	6	0	2	3	0	6
D,E	4	3	2	0	0	3	④
D,F	7	6	2	0	3	0	7
E,F	4	3	5	3	0	0	5

References

1. J.P Arabeyre, J. Fearnley, F.C. Steiger, W. Teathber, <u>Transportation Science, Vol.3</u>, pp 140-163 (1969).

2. O. Babic, D. Teodorovic, V. Tosic, <u>Aircraft Stand Assignment to Minimize Passenger Walking Distances</u>, 11th IFIP Conference on System Modeling and Optimization, Copenhagen, July 25-29 (1983).

3. R. H. Ballou, M. Chowdhurry, <u>The Logistics and Transportation Review, Vol. 16,</u> pp. 325-338 (1980).

4. E. J. Beltrami, L. D. Bodin, <u>Networks, Vol. 4,</u> pp. 65-94 (1974).

5. B. T. Bennett, R. B. Potts, <u>Vehicular Traffic Science</u>, American Elsevier Publishing Co., New York (1967).

6. O. Berman, R.C. Larson, A.R. Odoni, <u>European Journal of Operational Research, Vol.6,</u> pp. 104-116 (1981).

7. D. Blumenfeld, M. Landau, <u>Transportation Science, Vol. 6,</u> pp. 131-136 (1972).

8. L.D. Bodin, L. Berman, <u>Transportation Science, Vol. 13,</u> pp. 113-129 (1979).

9. L.D. Bodin, B. Golden, <u>Networks, Vol. 11</u>, pp. 97-108 (1981).

10. R. Bon, <u>European Journal of Operational Research, Vol. 1,</u> pp. 85-89 (1977).

11. M.G. Bradford, W.A. Kent, <u>Human Geography</u>, Oxford University Press, Oxford (1979).

12. R.L. Church, R.S. Garfinkel, <u>Transportation Science, Vol. 12</u>, pp. 107-118 (1978).

13. G. Clark, J. Wright, Operations Research, Vol. 1 pp. 568-581 (1964).

14. D. Cvetkovic, M. Milic, Teorija grafova i njene primene, Naucna Knjiga, Belgrade (1977).

15. E.W. Dijkstra, Numerische Mathematik 1, 269, 1959.

16. S.E. Elmaghraby, Some Network Models in Management Science Springer-Verlag, Berlin (1970).

17. M.L. Fisher, R. Jaikumar, Networks, Vol. 11, pp. 109-124 (1981).

18. L.R. Ford, D.R. Fulkerson, Flows in Networks, Princeton University Press, Princeton (1962).

19. L.R. Foulds, Transportation Research, Vol. 15, pp. 273-283 (1981).

20. R.W. Flyod, Communications of ACM, 5, 345(1962).

21. M. Friedman, Transportation Research, Vol. 12, pp. 305-308 (1978).

22. B. Gavish, P. Schweitzer, Transportation Science, Vol. 8, pp. 13-23.

23. B. Gillett, J. Johnson, ORSA/TIMS Meeting, San Juan Puerto Rico (1974).

24. B. Gillett, L. Miller, Operations Research, Vol. 22 (1974).

25. B.L. Golden, R.T. Wong, Networks, Vol. 11, pp. 305-315 (1981).

26. G. Gunawardane, European Journal of Operational Research, Vol. 10, pp. 190-195 (1982).

26[a] S.L. Hakimi, Operations Research, Vol. 12, pp. 450-459 (1964).

27. S.L. Hakimi, E.F. Schmeichel, J.G. Pierce, Transportation Science, Vol. 12, pp. 1-15 (1978).

REFERENCES

28. G.Y. Handler, Transportation Science, Vol. 7, pp. 287-293 (1973).

29. R.C. Larson, A.R. Odoni, Urban Operational Research, MIT Press, Cambridge, Mass (1981).

30. G. Laporte, Y. Nobert, European Journal of Operational Research, Vol. 6, pp. 224-226 (1981).

31. A. Levin, Transportation Science, Vol. 5, pp. 232-255 (1971).

32. T.L. Magnati, Networks, Vol. 11, pp. 179-213 (1981).

33. Lj. Martic, Nelinearno programiranje, Informator, Zagreb (1973).

34. J.W. Male, J.C. Liebman, C.S. Orloff, Networks, Vol. 7, pp. 89-92 (1977).

35. C. Mandl, Applied Network Optimization, Academic Press, London, New York (1979).

36. R.E. Marsten, F. Shepardson, Networks, Vol.11 pp. 165-177 (1981).

37. R.W. Matthaus, Transportation Science, Vol. 10, pp. 216-221 (1976).

38. P.B. Mirchandani, A.R. Odoni, Transportation Research, Vol. 13, pp. 113-122 (1979).

39. N. Muller-Merbach, Transportation Research, Vol. 8, pp. 377-378 (1976).

40. G.J. Murphy, Transport and Distribution, Business Books, London (1978).

41. G.F. Newell, Traffic Flow on Transportation Networks, MIT Press, Cambridge, Mass. (1980).

42. G.S. Orloff, Networks, Vol. 4, pp. 35-64 (1974).

43. C.S. Orloff, Networks, Vol. 4, pp. 147-162 (1974).

44. C.S. Orloff, D. Caprera, Transportation Science Vol.10, pp. 361-373 (1976).

45. R. Petrovic, <u>Specialne metode u optimizaciji sistems</u>, Tehnicka Knjiga, Belgrade (1977).

46. J.F. Pierce, <u>Transportation Research, Vol. 3</u>, pp. 1-42 (1969).

47. M.C. Poulton, A. Kanafani, <u>Transportation Science, Vol 9</u>, pp. 224-247 (1975).

48. O.M. Raft, <u>Journal of Operational Research, Vol 11</u>, pp. 67-76 (1982).

49. I. Rallis, <u>Intercity Transport</u>, The Macmillan Press Ltd, London (1977).

50. J. Rubin, <u>Transportation Science, Vol. 7</u>, pp. 34-48 (1973).

51. R. Russell, W. Igo, <u>Networks, Vol. 9</u>, pp. 1-17 (1979).

52. D. Schilling, D.J. Elzinga, J. Cohon, R. Church, C. ReVelle, <u>Transportation Science, Vol. 13</u>, pp. 163-175 (1979).

53. P.A. Steenbrink, <u>Optimization of Transport Networks</u>, John Wiley & Sons, London (1974).

53[a]. H.A. Taha, <u>Operations Research</u>, Macmillan Publishing Co. Inc., New York (1982).

54. D. Teodorovic, <u>Saobracaj</u> no. 2, pp. 313-316 (1983).

55. D. Teodorovic, S. Guberinic, <u>European Journal of Operational Research, Vol. 15</u>, no. 2 (1984).

56. F.A. Tillman, <u>Transportation Science, Vol. 3</u>, pp. 192-204 (1969).

57. G.P. White, <u>European Journal of Operational Research, Vol. 9</u>, pp. 190-193 (1982).

58. J.J. Wiorkowski, <u>Transportation Research, Vol. 9</u>, pp. 181-185 (1975).

Index

Absolute center	187,192
Acyclic oriented graph	156–158,161
Algorithm	10,11,13–23,27,30,32,35 37,41,44–46,57,72,74,75 77,80,118–121,124,130, 132,133,135,142,143, 147–149,151,186,192,195 196,198,200–202
Artificial branch	77,86
Artificial path	80,85
Assignment algorithm	142
Backtracking	97,98,101,102,106, 109–111,113
Basic variable	83
Beltrami	78,79
Bipartite graph	156,162,163,165–170
Blumenfeld	30
Bodin	78,79,125
Bounding	116,182

INDEX

Branch	1,2,4,5,7,11,12,14,16, 20–22,24,25,29,30,32–43 45,47,49,50,52–54,57–59 63,65,69,72–75,77,78,86, 90,119,122,126,129,133, 135–138,156,160–170, 174–176,181,186,187,189, 191,192,195
Branching	88,89,90–92,94–114,116, 179,182
Branch-and-bound	65,88,90,113,177–179,182
Branch capacity	56,57
Center	186–189,192,203
Centroid	50–52
Chain	5,57–63,156–158,159, 161–164,170,175–177,179, 182,183
Chinese postman problem	68,69,71,74,78–80,84
Clark	132,135,139,142,143,148, 149
Closed state	10,11,12–16,18,20
Combinatorial problem	66,67
Combinatorial programming	65,66,88,114,118
Connected subgraph	7,8

Conservation laws	50,52,53
Crew scheduling	65,66
Cut set	54,56,57
Cycle	5,7,8,42,45,69,70,156
Degree	4,73
Demand node	79,80
Depot	87,117,118,126,129–131, 133,134,139,140,142–145, 150–155,185,194,199,200
Deterministic network	35–37
Dijkstra	10,11,13,21
Discrete random variable	36
Dispatching strategy	172
Dynamic programming	65,66
Edge covering problem	65,68
Elementary path	5
Euler	69,70,78,79
Euler path	69–72
Euler tour	69,71,72,74,77,79,80,86, 119,122,123
Even degree node	72
Fisher	142,143,147–149
Fleet	126–128,132,172,179

Flow	49–54, 162–170
Flyod	21, 22
Gillett	139, 143, 148–151
Golden	125
Graph	1, 40, 156, 157, 162, 163, 170, 176, 181, 182
Guberinić	172
Hakimi	192, 194, 196, 198, 200
Heuristic procedure	65, 66, 88, 114
Immediate predecessor node	11, 12, 14–19, 21, 23, 24, 29
Indegree	4, 79–81
Intermediate node	50–52
Internal point	133, 135, 137, 138
Jaikumar	142, 143, 147–149
Johnson	150, 151
Landau	30, 32
Larson	77, 200
Level of service	1
Levin	155, 156
Linear programming	79, 80, 83
Little	92
Local center	186, 189–192

INDEX 215

Location	185,188,189,192,194,195, 198—202
Lower bound	91,96
Matrix reduction	92
Maximum flow	57,58,63,64
Median	194—196,198,200,201
Miller	139,143,148,149
Minimax problem	186
Minimum spanning tree	40—43,47,48,121,122
Mixed network	2,4,13
Müller-Merbach	67
Node	1,2,4—8,10—25,27,29—34,36, 39—43,45,47,51,52,54,56, 57,59,65,66,69,71,73,74, 75,77,79—81,85—92,94, 96—99,101—104,106,107, 109—112,114—119,121,123, 125,126,128,131,134,135, 142,143,145—148,151—157, 159—162,164—168,170, 174—177,179,181,186,189, 193—199,201,203—205
Node center	187
Node covering problem	65,86

Nonbasic variable	83
Nonoriented connected network	5,6,70—72
Nonoriented network	2—5,7—9,13,35,41,69, 72—75,79,187,192,193,195, 202
Objective function	67,127,201,202
Odd degree node	70—78,119,122
Odoni	77,200
Open state	10—13,15—17
Optimal solution	66,74,84,91,97,100,105, 106,108,112,114,130,155, 178,179,182,183
Oriented branch	51
Oriented connected network	5,6,79
Oriented graph	7
Oriented network	2—5,9,13,78—80,84,156,202
Oriented tree	8,9
Outdegree	4,79—81
Pairwise matching	75,77—79,119
Partial graph	7
Path	4—8,12,31,32,34,36,69,85, 88,90,91,93,94,97,99—101,

INDEX 217

	105,106,108
Permanent label	10
Permutation	87,88,115
Pierce	87,88
Polarity	79—81,85
Predecessor matrix	21—23
Probabilistic network	34—36,38,39
Probability	36—39
Probability distribution	35,38
Random variable	34,35,38
Reliable heuristic programming	88
Requirement problem	204
Root node	8,9
Route	65,69,87,114—118,127,128, 131—133,135,137—142, 145—148,151,153,154,159, 161,170,171,181
Routing	124,125,139,141,150
Saving	132,133,135—137
Schedule perturbation	172
Scheduling	124,125
Seed point	139—141
Shortest path	10—15,18—23,25,29,30,

218 INDEX

	34–37,39,69,74,77,79,80, 81,188,196,197,203
Shortest path lenght matrix	21,22
Simple path	5
Sink	53,54,56,57,164
Source	53,54,56,57,164
Space-time diagram	159,164,170,179–181
Spanning tree	8,9,40
Steady state flows	50
Strongly connected oriented network	5–7
Subgraph	7
Supply node	79,80
Sweeping algorithm	139–141,143
Tentative label	10
Teodorović	172
Total flow	53,54,56,170
Tour	118,120,123–125,128–131
Traffic	1,49,50,53,65,67,86,181
Transportation	1,53,65,68,86,114,115, 118,125,127,128,131,156, 193
Transportation network	1,2,10,11,13–18,21,24,29,

	30,32,33,40–42,44,46, 50–54,56–58,64–67,69,86, 91,114–116,118,125,126, 128,129,142,147,148,155, 158,161,172,174,176,185, 186,188,189,193,195,196, 198–203
Transportation network capacity	56
Transportation system	1,21,67,185
Traveling salesman problem	65,67,86,118,119,128–131, 142,147,177,179
Travel time	14,29,35,69,90–94,96,97, 99,100,102,103,105,112, 115,186
Tree	7,8,40,41,45,47,88–90,94, 96,97,113,114,179
Upper bound	97,100,105,106,108
Vehicle routing problem	65,86,125,126
Vehicle scheduling	66
Wright	132,135,139,142,143,148, 149

For Product Safety Concerns and Information please contact our EU
representative GPSR@taylorandfrancis.com
Taylor & Francis Verlag GmbH, Kaufingerstraße 24, 80331 München, Germany

www.ingramcontent.com/pod-product-compliance
Lightning Source LLC
Chambersburg PA
CBHW070604300426
44113CB00010B/1396